The New and Expanded

L & M in the U.S. Seagoing Services:

101 Basic Leadership and Management

Concepts for the Military Professional

Authored by

Kenneth R. Bryan

IMPORTANT NOTICE

U.S. Copyright has been applied for with the U.S. Copyright Office, 101 Independence Avenue S.E. Washington, DC 20559-6000.

Reference Case # 1-359157921 & 1-362445531

Table of Contents

SEAGOING SERVICE L & M PRINCIPLE (8):
It is not about ego, it is about mission.

SEAGOING SERVICE L & M PRINCIPLE (9):
The seagoing services are based on technology; an understanding of technical matters is important.

SEAGOING SERVICE L & M PRINCIPLE (10):
Creation of understanding has merit.

SEAGOING SERVICE L & M PRINCIPLE (11):
Do you own the problem?

SEAGOING SERVICE L & M PRINCIPLE (12):
Anticipate, recognize, evaluate, and control hazards.

SEAGOING SERVICE L & M PRINCIPLE (13):
It is important to have an understanding of the detailed requirements; understanding their intent also matters.

SEAGOING SERVICE L & M PRINCIPLE (14):
We are members of a team. We need to be supportive of the professional growth and development of each other and individual well-being.

SEAGOING SERVICE L & M PRINCIPLE (15):
Adherence to core values.

SEAGOING SERVICE L & M PRINCIPLE (16):
Seek to establish professional credibility.

SEAGOING SERVICE L & M PRINCIPLE (17):
What is the impact of your action? Inaction?

SEAGOING SERVICE L & M PRINCIPLE (18):
Communicate upward, laterally, and downward.

SEAGOING SERVICE L & M PRINCIPLE (19):
People deserve respect.

SEAGOING SERVICE L & M PRINCIPLE (20):
Do not be buffaloed, intimidated, or pressured to make the wrong decision.

SEAGOING SERVICE L & M PRINCIPLE (21):
We're always on-the-clock.

SEAGOING SERVICE L & M PRINCIPLE (22):
Integrity counts.

SEAGOING SERVICE L & M PRINCIPLE (23):
Honor commitments.

SEAGOING SERVICE L & M PRINCIPLE (24):
At some point you will be lied to. In your response, demonstrate maturity.

SEAGOING SERVICE L & M PRINCIPLE (25):
Just because you have the authority to take enforcement action doesn't mean that enforcement action is the most appropriate or effective direction to proceed in.

SEAGOING SERVICE L & M PRINCIPLE (26):
When individuals senior to you in your Chain-of-Command make a decision which may be counter to your desired outcome, seek to understand the factors which have influenced their decision making. At some point you may be standing in their shoes.

SEAGOING SERVICE L & M PRINCIPLE (27):
Those junior to you are the future of your service.

SEAGOING SERVICE L & M PRINCIPLE (28):
Be accountable.

SEAGOING SERVICE L & M PRINCIPLE (29):
When you've made an error, learn from it and move on.

SEAGOING SERVICE L & M PRINCIPLE (30):
If expectations are not communicated, most likely they will not be met.

SEAGOING SERVICE L & M PRINCIPLE (31):
Documentation counts. It helps the team down the road understand what happened and why decisions were made.

SEAGOING SERVICE L & M PRINCIPLE (32):
There is a time and place for very frank communications, and a time and place for more circumspect communications. Learn the difference.

SEAGOING SERVICE L & M PRINCIPLE (33):
Be supportive of the Command Cadre. They are more supportive of you than you most likely realize.

SEAGOING SERVICE L & M PRINCIPLE (34):
It is okay to admit that you're not sure, and that additional research is necessary.

SEAGOING SERVICE L & M PRINCIPLE (35):
Advance preparation promotes successful outcomes.

SEAGOING SERVICE L & M PRINCIPLE (36):
Be accessible.

SEAGOING SERVICE L & M PRINCIPLE (37):
Seek out training opportunities.

SEAGOING SERVICE L & M PRINCIPLE (38):
Support the training of your subordinates.

SEAGOING SERVICE L & M PRINCIPLE (39):
Maintain and lend reference materials.

SEAGOING SERVICE L & M PRINCIPLE (40):
Are proposed solutions operationally, technologically and economically feasible?

SEAGOING SERVICE L & M PRINCIPLE (41):
Maintain appropriate professional distance from your subordinates.

SEAGOING SERVICE L & M PRINCIPLE (42):
Expect professionalism, not perfection.

SEAGOING SERVICE L & M PRINCIPLE (43):
Stewardship.

SEAGOING SERVICE L & M PRINCIPLE (44):
Thoroughness adds value; however understand the concept of diminishing marginal returns.

SEAGOING SERVICE L & M PRINCIPLE (45):
Safety of life, property, and the marine environment.

SEAGOING SERVICE L & M PRINCIPLE (46):
Don't be afraid to get dirty. It is part of the job.

SEAGOING SERVICE L & M PRINCIPLE (47):
Be a positive change agent.

SEAGOING SERVICE L & M PRINCIPLE (48):
When a member has punitive action initiated against them, the impact extends to the member's family, the community, and your unit.

SEAGOING SERVICE L & M PRINCIPLE (49):
Prevention of the death, injury, accident, and environmental harm is always preferable to a reactive response after bad outcomes have occurred. But if they have occurred, seek to prevent the next bad outcome.

SEAGOING SERVICE L & M PRINCIPLE (50):
Test your equipment before you need to use it.

SEAGOING SERVICE L & M PRINCIPLE (51):
Understand the limitations of your personal protective equipment.

SEAGOING SERVICE L & M PRINCIPLE (52):
Maintain fiscal responsibility.

SEAGOING SERVICE L & M PRINCIPLE (53):
Pragmatism.

SEAGOING SERVICE L & M PRINCIPLE (54):
Be attentive to your external environment including personal security.

SEAGOING SERVICE L & M PRINCIPLE (55):
When overseas, respect the laws and culture of the host nation.

SEAGOING SERVICE L & M PRINCIPLE (56):
Innovate.

SEAGOING SERVICE L & M PRINCIPLE (57):
Shooting-from-the-hip is best left to the Westerns.

SEAGOING SERVICE L & M PRINCIPLE (58):
Understand the end goal, the desired outcome.

SEAGOING SERVICE L & M PRINCIPLE (59):
Wear your safety equipment.

SEAGOING SERVICE L & M PRINCIPLE (60):
Limit occupational exposures.

SEAGOING SERVICE L & M PRINCIPLE (61):
Measurement and quantification in terms of distance, volume, mass, thickness, time and other metrics results in precision and ability to duplicate.

SEAGOING SERVICE L & M PRINCIPLE (62):
Use of appropriate terminology conveys understanding, experience and professionalism.

SEAGOING SERVICE L & M PRINCIPLE (63):
Recognize trends.

SEAGOING SERVICE L & M PRINCIPLE (64):
When things go wrong, keep it all in perspective.

SEAGOING SERVICE L & M PRINCIPLE (65):
Create effective and appropriate efficiencies.

SEAGOING SERVICE L & M PRINCIPLE (66):
We have multiple customers.

SEAGOING SERVICE L & M PRINCIPLE (67):
Not all issues are of equal importance.

SEAGOING SERVICE L & M PRINCIPLE (68):
What is the root cause?

SEAGOING SERVICE L & M PRINCIPLE (69):
Carry up to date business cards.

SEAGOING SERVICE L & M PRINCIPLE (70):
Recognize the power centers.

SEAGOING SERVICE L & M PRINCIPLE (71):
A picture can speak a thousand words.

SEAGOING SERVICE L & M PRINCIPLE (72):
Don't be too accepting of industry hospitality.

SEAGOING SERVICE L & M PRINCIPLE (73):
From most effective to least effective, the hierarchy of health and safety controls are elimination or substitution, engineering controls, warnings, training & procedures & other administrative controls, and personal protective equipment.

SEAGOING SERVICE L & M PRINCIPLE (74):
Fatigue kills.

SEAGOING SERVICE L & M PRINCIPLE (75):
An operation takes as long as it needs to take.

SEAGOING SERVICE L & M PRINCIPLE (76):
Intoxication increases personal risk.

SEAGOING SERVICE L & M PRINCIPLE (77):
The Pocket Memo Book.

SEAGOING SERVICE L & M PRINCIPLE (78):
Walk-the-Talk.

SEAGOING SERVICE L & M PRINCIPLE (79):
RHIR, not RHIP.

SEAGOING SERVICE L & M PRINCIPLE (80):
Be sensitive to pay grade and positional
power differences.

SEAGOING SERVICE L & M PRINCIPLE (81):
Contribution.

SEAGOING SERVICE L & M PRINCIPLE (82):
Acknowledging human milestones.

SEAGOING SERVICE L & M PRINCIPLE (83):
The three dimensional employee.

SEAGOING SERVICE L & M PRINCIPLE (84):
Moving beyond disappointment.

SEAGOING SERVICE L & M PRINCIPLE (85):
Recognition.

SEAGOING SERVICE L & M PRINCIPLE (86):
Operating within the legal construct.

SEAGOING SERVICE L & M PRINCIPLE (87):
Understanding of Motivation Theories and
individual motivators is important.

SEAGOING SERVICE L & M PRINCIPLE (88):
Understanding of economies and diseconomies of scale.

SEAGOING SERVICE L & M PRINCIPLE (89):
Performance measurements.

SEAGOING SERVICE L & M PRINCIPLE (90):
Advertising of services.

SEAGOING SERVICE L & M PRINCIPLE (91):
Minimization of least desired outcomes.

SEAGOING SERVICE L & M PRINCIPLE (92):
But Sir...

SEAGOING SERVICE L & M PRINCIPLE (93):
Don't throw gasoline on the fire.

SEAGOING SERVICE L & M PRINCIPLE (94):
Strike while the iron is hot.

SEAGOING SERVICE L & M PRINCIPLE (95):
There is a time and place for yelling at people.

SEAGOING SERVICE L & M PRINCIPLE (96):
Consistency without rigidity.

SEAGOING SERVICE L & M PRINCIPLE (97):
Vision Statement.

SEAGOING SERVICE L & M PRINCIPLE (98):
Morale matters.

SEAGOING SERVICE L & M PRINCIPLE (99):
Fix the problem not the symptom.

SEAGOING SERVIC L & M PRINCIPLE (100):
Leaders manage and managers lead.

SEAGOING SERVICE L & M PRINCIPLE (101):
Be true to your own style.

CONCLUDING REMARKS

APPENDIX A – Declaration of Independence.

APPENDIX B – Bill of Rights.

Preface

Why another book on Leadership and management? There have been hundreds of them published, so what does one more add to the mix? Great questions!

This book on leadership and management is targeted to those in the U.S. military who serve as members of the seagoing services in any capacity. Within its pages are worthwhile recommendations for every service member from Seaman Recruit through and including Flag Officer.

Regardless of how much experience any individual has, they can always learn more. I count myself in this category. Although I have spent more than thirty years active duty in a seagoing service, and served in the enlisted ranks through Chief Petty Officer, in the Chief Warrant Officer ranks through CWO4, and in the Officer ranks through Commander, I'm still learning about leadership and management each and every day. Like some of you, I still get it wrong on occasion, and have to dust myself off and forge ahead, hopefully learning from my mistakes. This book is a good read. I hope you enjoy it, and learn a few things from the L & M (Leadership and Management) principles contained within its pages.

Best wishes for your success as a Leader, as a Manager and as a professional in a U.S. seagoing service.

Ken Bryan

Introduction

This text is designed for use in either a Leadership and Management Course, or as a work to be read by an individual external to a classroom environment.

The 'L & M Thought Exercises' lend themselves particularly well to group discussions. They will force you to think and draw conclusions based on your experiences and those of the other professionals in your group.

The leadership and management principles are purposely kept short. There is little value in taking ten pages to explain a principle which can be reasonably covered in on page.

IN ADVANCE OF THE SEAGOING SERVICE
L & M PRINCIPLES

Although the concepts of leadership and management are closely aligned, they are two distinct activities. If this book is being used as part of a Leadership and management course, it may prove advantageous to review the following 'L & M Thought Exercises' in advance of the Leadership and Management Principles described later in the text:

L & M Thought Exercise: Define leadership and define management. How are the two similar? How are they different?

L & M Thought Exercise: What are the duties and responsibilities of a Leader?

L & M Thought Exercise: What are the duties and responsibilities of a Manager?

L & M Thought Exercise: Which is more important, the exercise of good leadership or the exercise of good management?

SEAGOING SERVICE L & M PRINCIPLE (1):
We are privileged to be members of our service.

This is a principle worth remembering when we've had a particularly bad day aboard the ship, in the field, or in the office.

We are truly privileged to be able to serve the American public as members of the United States Military.

You are part of a long legacy of men and women which extends back into the past to Colonial times. Those before you have given of their spirits, and often flesh (life and limb), to help secure and maintain the freedoms we enjoy today. Next time your schedule permits, take the time to read the Declaration of Independence and U.S. Bill of Rights (Amendments 1 – 10 of the U.S. Constitution). Your fellow service members, past and present, have fought and died for the freedoms expressed in these documents which were drafted by the Founders. These two documents are contained in Appendices A and B.

L & M Thought Exercise: Name three reasons why you are privileged to be a member of your specific seagoing service.

L & M Thought Exercise: What can we do to honor those who served before us?

SEAGOING SERVICE L & M PRINCIPLE (2): The seagoing services are engaged in missions that are complex and ever evolving.

"We now accept the fact that learning is a lifelong process of keeping abreast of change. And the most pressing task is to teach people how to learn." - Peter Drucker

Military missions are complex and ever evolving. How does this relate to Leadership and management?

Yesterday's solutions may not be the solutions which are required today, particularly from a leadership and management perspective. As a Leader and Manager you need to stay current with the changes occurring in the seagoing services. Professional readings and participation in training opportunities help. Getting out in the field, where the steel meets the water, and meeting and working with those at the tip of the spear helps. Continuing to learn and seeking knowledge and understanding also promotes knowledge of the continuing evolution which is occurring.

L & M Thought Exercise: What have you done in the past three months to improve your ability to meet the complex missions your service is engaged in?

L & M Thought Exercise: Visualize your current operational specialty in five to ten years. What have you done to prepare tomorrow's Leaders and Managers for the changes you see coming?

SEAGOING SERVICE L & M PRINCIPLE (3): On-the-job safety matters.

According to OSHA and the Bureau of Labor Statistics, members of the resident military are significantly more likely to be fatality injured than non-military workers. In 2007, the fatal injury rate for members of the resident military was 5.5 fatalities per 100,000 workers compared to the all worker rate of 3.8. The leading event for fatal injuries to resident military is transportation incidents which accounted 6 out of 10 deaths. Of the 215 fatal injuries which occurred between 2003 and 2008, 127 (59 percent) were aircraft incidents; two-thirds of which involved helicopters. In regards to on-the-job safety, Leaders have the responsibility to promote safety; adhere to policies and procedures relating to safety, be attuned to the existence of 'best practices' to reduce and eliminate risk, and to incorporate those 'best practices', and assure that those under their command are provided with the tools and training necessary to work safely. Look out for the safety, health and well-being of subordinates, peers, and superiors.

L & M Thought Exercise: List three things you can readily do to promote and achieve a safer work environment for your subordinates.

L & M Thought Exercise: What is 'risk management' and how does it apply to your current assignment?

SEAGOING SERVICE L & M PRINCIPLE (4): Don't allow your staff to be marginalized.

As a Chief Warrant Officer, I can recall at one point where I was 'supervised' by a Lieutenant, Lieutenant Commander, and a Commander while running off photocopies of a public safety announcement for an event which was going to be attended by an estimated one hundred thousand people. The 1,000 public service announcements which were printed were going to be piggy-backed at a display booth being operated by another component of my military branch. It took a lot of convincing that the 1,000 public service single sheet, double sided, announcements which were being printed wasn't excessive (As it turns out, thousands more could and should have been distributed). It wasn't a good feeling to be marginalized by my Command on a very simple issue. To my chagrin, I've also fallen into the same trap of providing over direction and marginalizing the solid professionals on my team. If you've got good, capable, trained staff, be attentive to any natural inclinations you may have to over-supervise, over-direct, and professionally marginalize your staff.

L & M Thought Exercise: What can you do as a Leader and Manager to prevent your staff from being marginalized?

L & M Thought Exercise: How should you address the situation when external parties marginalize your staff?

SEAGOING SERVICE L & M PRINCIPLE (5):
Be on time.

Being on time for a meeting shows respect for the schedules of those you are meeting with and also personal preparedness.

If you have a meeting of six people, and one key participant is late by ten minutes, your organization hasn't lost ten minutes of productivity. An hour has been lost (10 minutes x 6 individuals).

If you get in the habit of respecting people's schedules, they will learn to respect yours.

Meeting professional submission times also matters. Often your work is one small component of a larger project. The delay in one component potentially creates a ripple effect which can derail output.

In a professional organization, timeliness does matter.

L & M Thought Exercise: Think about your pending meetings and deadlines. If you believe that the chance exists that you won't be on time or meet an established deadline, what is within your span of control to correct the situation?

L & M Thought Exercise: For meetings which you chair, what is the best means to promote timeliness?

SEAGOING SERVICE L & M PRINCIPLE (6): Off-the-job safety matters.

According to the National Safety Council, annually about 180,000 Americans die from off-the-job injuries with motor vehicle and home related deaths topping the list.

Annually, millions more Americans sustain serious injuries off-the-job.

Big contributors are fatigue, use of alcohol and other intoxicants, inexperience, inadequate training, and other human factors.

So how do you promote off-the-job safety? Essentially, the same way one promotes on-the-job safety...

L & M Thought Exercise: Think about the riskiest behavior you engage in off-the-job. What can you do to better manage or eliminate the risk associated with the activity?

L & M Thought Exercise: Think about the riskiest behaviors that members of your staff engage in off-the-job. What can you do to better manage or eliminate the risk associated with their activities?

SEAGOING SERVICE L & M PRINCIPLE (7): Most service members and their civilian counterparts are solid professionals committed to doing the right thing.

Since the completion of obligated service by those members of the military who were drafted (involuntarily conscripted) in the 1970's, your co-workers have volunteered to be in the U.S. military. By and large they are good solid professionals committed to doing the right thing. As a Leader and Manager, this principle is important in that on occasion you are going to have to address the conduct or performance of an individual whose activities in these two arenas are outliers. In other words, they are simply not adhering to the standards and expectations of your organization and you as their supervisor.

A few things to think about:

Is the failure an individual failure, or is it a systems failure which impacts more than one person?

If you are considering taking action which could adversely impact the individual's career, understanding where the failure lies is essential. Was the person properly trained before they were assigned the task? Were expectations conveyed to the individual? Was the individual provided with the proper equipment to conduct the task? Was the individual set up for success? Or, was he or she set up for failure?

You catch the drift. If the professionals on your team are failing, then it is worth looking at factors which may extend beyond the individual.

L & M Thought Exercise: Do you agree with the statement that most service members and their civilian counterparts are solid professionals committed to doing the right thing? Why, or why not?

L & M Thought Exercise: Summarize your leadership and management responsibilities to the professionals in your seagoing services.

SEAGOING SERVICE L & M PRINCIPLE (8)
It is not about ego, it is about mission.

Ego is essentially an individual's idea of his or her own importance or worth. Generally, it is usually of an appropriate level.

Sometimes, we, as humans, let our egos get in the way of correct decision making. Because of our feelings of being slighted, or not given the appropriate level of recognition, we personalize the situation instead of making leadership and management decisions based on the facts and particulars of the situation.

Have you ever heard the following from a shipmate?

"Heck with him, he's a XXXXXXXX. I'll be darned if I'm going to provide him the support he's asking for."

If so, perhaps it is ego doing the talking instead of the rational, committed Leader and Manager. It is not about ego, it is about mission.

L & M Thought Exercise: Can you think of a situation where you let ego get in the way of mission? How could you have prevented this from happening?

L & M Thought Exercise: When there is a pre-existing deteriorated or non-existent working relationship because of egos, how can you move from where you are at, to where you need to be?

SEAGOING SERVICE L & M PRINCIPLE (9): The seagoing services are based on technology; an understanding of technical matters is important.

Every year our respective seagoing service gets more complicated and more dependent on technology.

Recently as an Acting Executive Officer I attended meeting filled with a dozen Information Technology types and one non-IT type. Guess who was that person?

Did I wish I had a better understanding of the issues being discussed? Of course.

As much as you can, it helps to understand the technology that your division, department, and unit are routinely involved with. This understanding will make you more effective and better able to address important issues.

L & M Thought Exercise: What is the most important technology your team is directly involved with? How do you improve your understanding of it?

L & M Thought Exercise: How can you improve your staff's understanding of relevant workplace technological issues?

SEAGOING SERVICE L & M PRINCIPLE (10): Creation of understanding has merit.

In the fleet there exists a commentary on the experience level of many O-1s which includes Ensigns and Second Lieutenants.

Question: "How do you destroy the career of an Ensign (or Second Lieutenant)?

Answer: "Do exactly what they tell you to do!"

The point to this is that as Leaders and Managers (including the above mentioned Ensigns and Second Lieutenants), we often want our staffs to meet the intent of what we desire, rather than the precise specifics of what we order.

This requires the Leader and Manager to spend a little bit of time to communicate 'Command Intent.' Try it - the understanding you create will serve you well. When your team understands your vision, goals, and your priorities you are halfway on the road to success.

L & M Thought Exercise: What issues don't your subordinates understand which they should understand? What are you doing to correct this?

L & M Thought Exercise: Do your subordinates have access to vision statements, goals, objectives and operational doctrines which serve to create understanding? If not, why not?

SEAGOING SERVICE L & M PRINCIPLE (11): Do you own the problem?

It is a good question to ask yourself if you are actually the owner of the problem and it is within your span of control to fix, or if the problem belongs to someone else, and you need to notify them of it so they can address it.

If you do 'own' the problem, then fix it. Many experts have found the below problem solving technique to be helpful:

Step 1: Problem identification.
Step 2: Alternatives generation.
Step 3: Evaluation of alternatives.
Step 4: Decision making.
Step 5: Implementation of the solution.
Step 6: Follow-up evaluation.

If the problem spans multiple jurisdictions, consider demonstrating the initiative you're capable of and spearhead the improvement effort!

L & M Thought Exercise: Think about a real life example in which you had to deal with the issue of jurisdiction. How did you address it?

L & M Thought Exercise: If you aren't sure if you own the problem, how do you figure out if it is within the scope of your responsibilities?

SEAGOING SERVICE L & M PRINCIPLE (12): Anticipate, recognize, evaluate, and control hazards.

If you are an Industrial Hygienist, you invariably will recognize this phrase. If you are not, it essentially means that you, as a Leader and Manager, should through experience, education, training, and judgment be able to reasonably anticipate hazards which might be present.

For example, if you are responsible for workers who go into confined spaces, you should be able to anticipate the potential hazards of the confined space. These hazards could include oxygen enrichment; oxygen depletion; the presence of toxic gases; the presence of combustible or explosive gases; routine hazards such as slips, trips, and fall hazards; electrical hazards; entrainment or engulfment hazards; hazards of being accidently sealed into the confined space if someone sealed the access; and so, on and so forth. Recognition involves being able to identify the actual hazards of the space. For example, if there are electrical wires or circuitry inside the space there could exist an electrical hazard until proper electrical tag-out occurs and is verified. Evaluation involves assessing the risk which is posed by recognized hazards. Control involves the actions undertaken to eliminate or reduce the threat of death, injury, and other undesired outcomes.

L & M Thought Exercise: Describe two potential hazards your team may potentially be exposed to, and think about how they can be better controlled.

SEAGOING SERVICE L & M PRINCIPLE (13): It is important to have an understanding of the detailed requirements; understanding their intent also matters.

Within my service there are tens of thousands of pages of guidance, instructions, and policies. Most of these, in some way, shape or form are relevant to my particular unit. Most likely, there are tens of thousands of pages of policies which are relevant to your unit.

As shocking as it may seem, despite my due diligence, I'm not totally familiar with every single line on every single one of the policies...no human could be. Most likely you are not either.

So while it is important to having as much of an understanding of the details of the guidance, instructions, and policies, it is also important to understand their intent so that you can refer to them should you need to. This holds true for areas beyond policy as well...

L & M Thought Exercise: What is the intent of the guidance, instructions, and policies relating to your department's core missions?

L & M Thought Exercise: Is there a publication within your seagoing service which provides a comprehensive listing of the current nationally promulgated policies and manuals from your organization?

SEAGOING SERVICE L & M PRINCIPLE (14): We are members of a team. We need to be supportive of the professional growth and development of each other and individual well-being.

Regardless of position in the organization, we are members of a team and are dependent on our team members.

Remember this! Repeat - Regardless of position in the organization, we are members of a team and are dependent on our team members.

Our team members can either support us, or give the appearance of support while engaging in activities counter to our interest, or be outright non-supportive and obstructionist.

By being supportive of the professional growth and development of each other, and individual well-being we help the team meet expectations and mission. We also help ourselves – giving support results in support being given to you.

L & M Thought Exercise: What steps are within your span of control to improve the professional growth and development of your team?

L & M Thought Exercise: For individuals and families with special or unique needs, what programs exist to provide assistance in your seagoing service?

SEAGOING SERVICE L & M PRINCIPLE (15): Adherence to core values.

"Live in such a way that you would not be ashamed to sell your parrot to the town gossip." - Will Rogers

Core values are an essential part of leadership and management in the seagoing services in that they serve to guide each of us in our conduct and supervision of others, particularly in the lack of defined policies or regulations. If your decision is consistent with the core values, in the absence of concrete policy, most likely it will be considered to be appropriate.

The core values of the United States Navy, as extracted without change from their official internet web site, are as follows:

Honor: "I will bear true faith and allegiance ..." Accordingly, we will: Conduct ourselves in the highest ethical manner in all relationships with peers, superiors and subordinates; Be honest and truthful in our dealings with each other, and with those outside the Navy; Be willing to make honest recommendations and accept those of junior personnel; Encourage new ideas and deliver the bad news, even when it is unpopular; Abide by an uncompromising code of integrity, taking responsibility for our actions and keeping our word; Fulfill or exceed our legal and ethical responsibilities in our public and personal lives twenty-four hours a day. Illegal or improper behavior or even the appearance of such behavior will not be tolerated. We are accountable for our professional and personal behavior. We will be mindful of the privilege to serve our fellow Americans.

Courage: "I will support and defend ..." Accordingly, we will have: courage to meet the demands of our profession and the mission when it is hazardous, demanding, or otherwise difficult; Make decisions in the best interest of the navy and the nation, without regard to personal consequences; Meet these challenges while adhering to a higher standard of personal conduct and decency; Be loyal to our nation, ensuring the resources entrusted to us are used in an honest, careful, and efficient way. Courage is the value that gives us the moral and mental strength to do what is right, even in the face of personal or professional adversity.

Commitment: "I will obey the orders ..." Accordingly, we will: Demand respect up and down the chain of command; Care for the safety, professional, personal and spiritual well-being of our people; Show respect toward all people without regard to race, religion, or gender; Treat each individual with human dignity; Be committed to positive change and constant improvement; Exhibit the highest degree of moral character, technical excellence, quality and competence in what we have been trained to do. The day-to-day duty of every Navy man and woman is to work together as a team to improve the quality of our work, our people and ourselves.

The U.S. Marine Corps also maintains the same core values of honor, courage and commitment, although they are explained and discussed in a warrior context as it relates to their service.

The core values of the United States Coast Guard, as also extracted without change from the service's official Leadership and Professional Development internet web site, are as follows:

Honor: Integrity is our standard. We demonstrate uncompromising ethical conduct and moral behavior in all of our personal actions. We are loyal and accountable to the public trust.

Respect: We value our diverse work force. We treat each other with fairness, dignity, and compassion. We encourage individual opportunity and growth. We encourage creativity through empowerment. We work as a team.

Devotion to Duty: We are professionals, military and civilian, who seek responsibility, accept accountability, and are committed to the successful achievement of our organizational goals. We exist to serve. We serve with pride.

The core values of your seagoing service are words to live by in the performance of your professional duties and your off-hours activities.

L & M Thought Exercise: Provide an example where use of the core values have helped you address a professional situation in the absence of clear policy.

SEAGOING SERVICE L & M PRINCIPLE (16): Seek to establish professional credibility.

Why seek professional credibility? Because if you possess it, it will allow you to accomplish missions and routine activities more efficiently and effectively.

Professional credibility is comprised of the components of trustworthiness and expertise. Trustworthiness is based on subjective factors such reliability, honesty, integrity, reputation. Expertise involves credentials, certification, information and history of performance (getting things done).

Through your daily actions and communications, you are the owner of your professional credibility.

L & M Thought Exercise: Think about the personal and professional qualities of an individual you believed exhibited 'professional credibility.' Verbalize the qualities which you observed.

L & M Thought Exercise: Think about someone who through their own actions or inactions lost their professional credibility. How could the situation have been prevented?

L & M Thought Exercise: In the past thirty days what have you done to improve your professional credibility?

SEAGOING SERVICE L & M PRINCIPLE (17): What is the impact of your action? Inaction?

Every decision we make, or forego, has an impact, an outcome. The effective Leader and Manager considers the outcomes of their actions and inaction.

It has been reported that the lookouts aboard the RMS Titanic did not have binoculars to use when they stood their watch, and that the vessel was constructed of metal which was particularly brittle and subject to fracturing in cold water environments.

If the watch supervisor of the lookouts and the designer of the Titanic knew the impacts of the decisions (action) relating to the binoculars and hull plating, would they have made the same decisions? Of course not.

If we think ahead about the consequences of our decision making we promote optimal decisions.

L & M Thought Exercise: Think about a situation where you took action and should have held the action in abeyance. What about a situation where you should have acted and remained inactive?

L & M Thought Exercise: Under what circumstance is it better to act than to defer action? When should you defer action?

SEAGOING SERVICE L & M PRINCIPLE (18): Communicate upward, laterally, and downward.

"Good communication does not mean that you have to speak in perfectly formed sentences and paragraphs. It isn't about slickness. Simple and clear go a long way." - John Kotter

Upward communication is that which we provide to those in our organization who are higher in the hierarchy, lateral communication is to those at the same level within our organization, and downward is to those in the hierarchy beneath us.

Communication creates understanding and commonality of purpose. It promotes mission accomplishment, appropriate management of risks, efficient and effective use of resources. It eliminates misunderstandings and vagueness.

L & M Thought Exercise: How can you improve your upward, lateral, and downward communication?

L & M Thought Exercise: What are the formal and informal mechanisms within your organization to facilitate communication?

L & M Thought Exercise: Is it possible to over communicate in the professional environment?

SEAGOING SERVICE L & M PRINCIPLE (19): People deserve respect.

"If you want to see the true measure of a man, watch how he treats his inferiors, not his equals."- J. K. Rowling

"Do right. Do your best. Treat others as you want to be treated." - Lou Holtz

The concept of respect relates to courteous regard for a person's feelings. It involves treating them as a 'human' without prejudice and bias, and with recognition that they contribute to the organization. The concept of respect goes beyond that which is traditional in the military towards those senior to us. It recognizes that each individual, without regard to pay grade, has merit as both a human being and a member of the team. It is worth remembering several things: Your subordinates possess unique and important skills that most likely you do not possess; The efficient and effective running of your unit is dependent upon those unique and important skills; and, Pay grade is a very transitory thing, as evidenced by the 25th Chief of Naval Operations, Admiral Jeremy Michael Boorda who rose from E-1 to E-6, and then from O-1 to O-10.

L & M Thought Exercise: Think about the last few times when you thought that you were treated with less respect than you deserved. Have you ever done the same to other people?

> **SEAGOING SERVICE L & M PRINCIPLE (20): Do not be buffaloed, intimidated, or pressured to make the wrong decision.**

"Courage is what it takes to stand up and speak; courage is also what it takes to sit down and listen." - Winston Churchill

In situations where group decision making is necessary, often individuals with the group feel buffaloed, intimidated or pressured into making the wrong decision. This can occur when members of the group avoid promoting viewpoints outside the comfort zone of consensus thinking, feel pressure because of rank structures or positional power, have the desire to avoid being seen as foolish, or prefer to avoid embarrassing issues or angering other members of the group. Ideas and decisions must critically tested, analyzed, and evaluated in the seagoing services. This takes courage. You have been included in the decision making process because your opinions and understanding of the situation have merit. Communicate your concerns appropriately and professionally.

L & M Thought Exercise: What is a good technique for you, as a Leader and Manager, to prevent yourself from being buffaloed, intimidates or pressured into making the wrong decision.

L & M Thought Exercise: What is 'group think'?

SEAGOING SERVICE L & M PRINCIPLE (21): We're always on-the-clock.

As a Leader and Manager it is important to remember that we are always on the clock.

24/7 our conduct and activities off the job reflect on our military service. It can reflect positively or negatively.

Civil infractions, indebtedness, even overly aggressive driving in a vehicle which is adorned with military logos proudly advertising your affiliation, and unnecessary confrontations reflect on the military and your specific branch.

Secondly, the term "we're always on the clock" means that you are always subject to recall. It happens. You have chosen a profession which isn't nine to five. Be thankful for the opportunity to serve when the after hour's call or mission arises.

You will not regret being a good representative of your seagoing service off the job.

L & M Thought Exercise: From your perspective is the statement 'We're always on-the-clock' true?

L & M Thought Exercise: Do Leaders and Managers within your seagoing service spend a lot of time and effort to address off-the-job issues? Please explain.

SEAGOING SERVICE L & M PRINCIPLE (22): Integrity counts.

"A good name, like good will, is got by many actions and lost by one." - Lord Jeffrey

Integrity involves adherence to moral principles, reputability, trustworthiness, righteousness, honesty and the core values.

At some point in your career, the perception of your character by others is going to be of vital significance. You cannot predict when this will occur, however there is a near certainty that it will happen.

How do you establish your reputation for integrity?

Consistency of appropriate conduct and action, and adherence to your organization's standards and core values.

L & M Thought Exercise: Remember a time when all you had in your favor was another individual's perspective on your integrity. How important was your personal and professional reputation in regards to integrity during this situation?

L & M Thought Exercise: Think of someone you know who has a reputation for lack of integrity. What did the individual do (or fail to do) which has resulted in his or her poor reputation?

SEAGOING SERVICE L & M PRINCIPLE (23): Honor commitments.

Sometimes we make commitments, which in retrospect we regret. Honoring commitments relates to your personal integrity, and also sends the very real positive message to the person or group you've made the commitment to regarding their place in your list of priorities.

If you've made the commitment, and can reasonably meet it, honor it. Meet the intent and specifics of your promise to others.

If you find yourself faced with not meeting a lot of your commitments, ask yourself 'why?' and then take corrective action to reduce recurrence.

L & M Thought Exercise: When you are unable to honor your commitments, what is the best way to forge ahead?

L & M Thought Exercise: When someone made a commitment to you and failed to meet it, how did the situation impact your future professional or personal relationship?

SEAGOING SERVICE L & M PRINCIPLE (24): At some point you will be lied to. In your response, demonstrate maturity.

"In a large organization it is frequently difficult for a Leader to have an accurate sense of the company's affairs because reports are often afraid to be candid about problems." - Mihaly Csikszentmihalyi

"Truthfulness is the main element of character." - Brian Tracy

It isn't pleasant being lied to. Yet, as a Leader and Manager you will be lied to by a small percentage of people.

A lie, if in the form of an official statement is a violation of Title 18 United States Code 1001 and the Uniform Code of Military Justice (UCMJ).

You have a choice to make in how to respond. In your response, take a few seconds to think on the issue. Is the person deliberately attempting to deceive you or misrepresent the facts? Or, is the person relaying information they believe to be true but is actually erroneous. Intent matters. Obviously you want no recurrence of the lie. You want or need to know the truth. Do you have any other objectives?

L & M Thought Exercise: What are some mechanisms an individual can use to address the issue of being lied to in a professional environment?

SEAGOING SERVICE L & M PRINCIPLE (25): Just because you have the authority to take enforcement action doesn't mean that enforcement action is the most appropriate or effective direction to proceed in.

"I have always found that mercy bears richer fruits than strict justice." – Abraham Lincoln

Leadership and management requires the use of discretion based on the facts, circumstance, and context of the situation.

The use of enforcement action comes at a cost to Commands in terms of personnel, administrative and other costs. Enforcement action comes at a cost to the person it is being administered against as well as their family.

Remember, in many cases there does exist discretion. Use it prudently and appropriately.

L & M Thought Exercise: Consider a situation in which you had the discretion to take a range of enforcement actions. What factors did you consider in deciding the most appropriate course of action?

L & M Thought Exercise: Should every violation of the UCMJ result in an Article 15 proceeding or a court martial?

SEAGOING SERVICE L & M PRINCIPLE (26): When individuals senior to you in your Chain-of-Command make a decision which may be counter to your desired outcome, seek to understand the factors which have influenced their decision making. At some point you may be standing in their shoes.

As a careerist in a seagoing service, I've had the opportunity to serve in pay grades E-1 through Chief Petty Officer (E-3 through E-5 twice as a result of broken service), Warrant Officer-One through Chief Warrant Officer-Four, and then from Lieutenant (O-3) through and including Commander (O-5).

What I've learned is that there are often considerations which those in the Chain-of-Command must address that may not be readily apparent to those in more junior positions. Commands are forced to address macro-organizational and external issues which may be beyond the scope and experience of more junior members. In addressing personnel issues, there are career survivability and retention issues, continuity of operations issues, good order and discipline concerns and many other considerations. If you seek to understand the multiple issues which must be considered in many decisions, it will help you as you proceed through the ranks and are forced to deal with similar issues. Learning from those around you, as they make decisions good and bad, will assure your future success.

L & M Thought Exercise: Describe some of the issues which Commands must consider in decision making which extend beyond those of department heads and divisions.

SEAGOING SERVICE L & M PRINCIPLE (27): Those junior to you are the future of your service.

"It takes Leaders to grow other Leaders."- Ray Blunt

If you ask a hundred individuals in any given seagoing service who the most important person in their service is, the most likely response that will you receive will be the person who holds the reigns in the organization. The 'top dog' to use common slang, the man (or woman) with the most stars or most brass on their shoulder boards. To some extent, your shipmates are correct. The CNO, the Commandant of the Marine Corps, and the Commandant of the Coast Guard are extremely important people. They guide their respective services for a period, usually four years, and ensure that the exigent crises' of the day are appropriately addressed. It is my belief, that the most important people in the military are the 'up and comers.' The future of your service, regardless if it is the Navy, Marine Corps, or Coast Guard rests with Seaman Recruit Gooblotz who is in the middle of basic training, or his or her cousin Cadet Gooblotz who is attending the Naval Academy or Coast Guard Academy, or is entering the military through another means of accession. As old hands, it is our responsibility to teach, guide, ensure enculturation, and mentor these members. It is an enormous responsibility....but as a solid professional, you are up to meeting the task.

L & M Thought Exercise: If you could teach a subordinate three things about leadership and management, what would those three things be?

L & M Thought Exercise: Are you engaged in the mentoring process? If no, why not?

SEAGOING SERVICE L & M PRINCIPLE (28): Be accountable.

Accountability is the acknowledgment and assumption of responsibility for actions, decisions, and outcomes.

The best advice I ever received in the military was as a 19 year old from a salty O-6, who was the Commanding Officer on the cutter I was serving aboard as an Oceanographic Technician. One day, while doing oxygen titrations to determine the amount of oxygen in the seawater column down to about 5,000 meters, I screwed it up. On the draft message to our Oceanographic Institute, I gave an excuse which was, in all honesty, nonsense. The Captain, as he was preparing to sign the draft message, read it.

In a gravelly voice he said *"Son, if you frigged up, tell 'em you frigged up."* My skipper crossed out my draft message and told the truth.

It was a great lesson to learn – be accountable! Good or bad, accept the responsibility for actions, decisions, and outcomes.

Being accountable is linked to other principles discussed in this book including adherence with core values and professional presence.

L & M Thought Exercise: What is the best means to hold ourselves and others accountable for their actions?

SEAGOING SERVICE L & M PRINCIPLE (29): When you've made an error, learn from it and move on.

"The only man who makes no mistakes is the man who never does anything." - Theodore Roosevelt

"Learning from mistakes and constantly improving products is a key in all successful companies." - Bill Gates

Humans make mistakes. We are embarrassed by them, sometimes frustrated by them, and we nearly always regret them.

When they happen, seek to minimize the resulting damage to internal and external parties which has occurred as a result of the error. Learn what you should and need to from the error, and move on. Take reasonable efforts to prevent the same thing from recurring in the future.

L & M Thought Exercise: When you as a Leader and Manager make an error which detrimentally impacts others, do you acknowledge the error and seek to minimize the harm which may have been created?

L & M Thought Exercise: How do you promote learning amongst your staff when they have made an error?

SEAGOING SERVICE L & M PRINCIPLE (30):
If expectations are not communicated, most likely they will not be met.

How many of you have ever experienced frustration in the work place because expectations were not clearly conveyed. More than we, as Leaders and Managers, would like to admit.

Communication of expectations facilitates commonality of vision, properly directed effort, reduction in rework, ability to measure and document work and mission performance.

Ambiguity kills adherence of output to un-communicated expectations.

L & M Thought Exercise: What is the best means by which expectations can be communicated?

L & M Thought Exercise: Provide a summary of the expectations which should routinely be communicated? (For example, one would be the project or tasking due date.)

L & M Thought Exercise: Have you communicated your expectations to your subordinates?

SEAGOING SERVICE L & M PRINCIPLE (31): Documentation counts. It helps the team down the road understand what happened and why decisions were made.

Most of us have run into professional situations where we just scratch our heads and wonder 'How the heck could our predecessors have made such a bone-headed decision on the issue?'

For the most part, our predecessors were smart, capable and rational people. They made decisions based on the facts that they were aware of and the context in which the decision had to be made.

When major decisions are made, it is worth documenting why they were made. You will not regret your decision to do so. Communicate with those who follow in your path...

L & M Thought Exercise: How much documentation is enough?

L & M Thought Exercise: What factual information should be contained within the documentation?

SEAGOING SERVICE L & M PRINCIPLE (32): There is a time and place for very frank communications, and a time and place for more circumspect communications. Learn the difference.

"Think twice before you speak, because your words and influence will plant the seed of either success or failure in the mind of another." - Napoleon Hill

Frank communications can cut through the crap so to speak and be instrumental in moving forward to getting things done.

By their very nature, frank communications can be painful, cause angst, and if incorrectly done cause significant harm.

Three thoughts: Often best held in a closed door session, preferably between only two people; Make sure you don't get your ego involved (See Principle 8); and, in very frank communications, demonstrate as much respect as you can for people and circumstances. It takes a lot longer to build bridges than it takes to tear them down.

L & M Thought Exercise: Recall a situation you were involved with which required frank communications. What were some of the factors which made you opt for frank communications rather than more circumspect communications?

As a junior enlisted member in the seagoing services I failed to understand how much effort was expended to support individuals by the Command Cadre. More than thirty years later as a member of a Command Cadre, I'm now aware of how much effort is actually undertaken.

Opportunities are offered which may not necessarily provide significant benefit to the unit, but are afforded as part of the overall professional growth and development of the individual.

Promotions are supported, communications are undertaken with detailers and prospective commands, and consideration is given to those seeking off-hours education.

When individuals have professional issues, there is significant time spent 'behind closed doors' to determine what the best means is to get the individual back on track.

Personal problems receive the same care and attention. Your Command Cadre wants you to succeed and expends significant effort to help assure your success. Provide your support to them.

L & M Thought Exercise: What are some of the ways you can support your Command Cadre?

SEAGOING SERVICE L & M PRINCIPLE (34):
It is okay to admit that you're not sure, and that additional research is necessary.

As an E-5 technician, I had about 90 percent of the answers to any given problem I encountered on the job. More than thirty years later, after earning two undergraduate degrees and four master's degrees, and serving at fourteen primary duty stations in fourteen pay grades, I'm at the point where I have about 45% of the answers, my team has another 45% and the remaining 10% of the answers are held by people external to the immediate team and I.

No one expects you to have all of the answers, all of the time, at a moment's notice. Nobody is that good. Not even the Commander in Chief. Not even the best Admirals and Generals. Think about it...

L & M Thought Exercise: When you are not sure of something, how do you forge ahead?

L & M Thought Exercise: Often when individuals are not certain of the direction they should proceed in, they examine existing policy to obtain guidance. Is there a document within your seagoing service which provides a listing of the current existing policies/instructions? If so, examine this document.

SEAGOING SERVICE L & M PRINCIPLE (35): Advance preparation promotes successful outcomes.

The D-Day Invasion of June 6, 1944, in which nearly 160,000 troops crossed the English Channel for the Normandy, France landings, had been planned for since 1942. The initial draft of the invasion plan was approved in August 1943, ten months prior to the landings.

The advance preparations which took place assured a successful operation.

It hoped that you will never be involved in planning an operation as logistically complex and as critical as D-Day. However, it is important to remember that even far less significant operations require advance planning.

Something as simple as refueling your ship needs to be planned in advance if you want a high confidence level that the operation will be successfully completed without issues.

L & M Thought Exercise: Recall a situation when you did not prepare. Did an optimal outcome result?

L & M Thought Exercise: What contingency plans are in existence which relate to your operations. Are they current? Have you read them?

SEAGOING SERVICE L & M PRINCIPLE (36): Be accessible.

In the seagoing services, all individuals have certain responsibilities and authorities. Some individuals are amongst a limited few who can approve or engage in certain actions.

For example, for purchasing goods and services above a certain amount of money, most likely a contracting officer with special authority (and responsibility) needs to approve the expenditure.

If he or she is not accessible, either the expense isn't made and operations are potentially detrimentally impacted, or the system ignores its own procedures and 'shoots from the hip' (See Principle 57). Invariably this results in sub-optimization of outcomes.

Be accessible, and if you cannot be, delegate the authority if permitted under service regulations.

L & M Thought Exercise: Can a person be too accessible?

L & M Thought Exercise: What can those in leadership and management positions do to promote improved accessibility?

SEAGOING SERVICE L & M PRINCIPLE (37): Seek out training opportunities.

As a Leader and Manager, consistently seek out training and professional growth & development opportunities.

If you obtain training, you are increasing your value to your seagoing service and future employers. You are also enhancing your capacity to meet your responsibilities thereby reducing stress, rework, and errors.

If you seek training and are unable to obtain it, it is still a winning situation for you, in that you've established a willingness and desire to improve your professional skill sets. Most likely this will improve your opportunity to receive the training in the future.

L & M Thought Exercise: What is the difference between training and education?

L & M Thought Exercise: Within your seagoing service, where can you find information regarding training opportunities?

L & M Thought Exercise: What are some additional advantages, beyond those already noted, of having a trained work force?

SEAGOING SERVICE L & M PRINCIPLE (38): Support the training of your subordinates.

"As skills improve, one is able to take on greater challenges."
- Mihaly Csikszentmihalyi

Mission accomplishment requires that your subordinates possess the knowledge and skills necessary for them to perform their duties.

There are many advocates for just-in-time training. Just-in-time training permits the trainee to meet task specific demands with current knowledge.

There is only one catch to this concept. Sometimes it is not possible to predict when the skills will be needed. Absent the ability to predict when the training will be needed, using this principle will result in your staff being potentially unable to address unplanned for, unanticipated events.

Training of your team increases the team's ability to meet surge demands and unusual events. A trained staff provides you with greater flexibility in decision making.

L & M Thought Exercise: When should you not support the training of your subordinates?

L & M Thought Exercise: Within your organization what policies relate to training and professional growth and development?

SEAGOING SERVICE L & M PRINCIPLE (39): Maintain and lend reference materials (if unclassified, and permitted to be lent).

"Know where to find the information and how to use it. That's the secret of success." - Albert Einstein

This practice improves the overall ability of the team to meet mission. By being helpful to others, you garner their support in the future. At some time, the support of others will determine your success or failure.

Your team is like an anchor chain, with each member being a link. Strengthening links through lending of reference materials is appropriate.

L & M Thought Exercise: With the advent of the electronic and internet age where many reference materials are available electronically, which ones should be maintained in hard copy?

L & M Thought Exercise: Do you maintain and lend reference materials? Why, or why not?

SEAGOING SERVICE L & M PRINCIPLE (40): Are proposed solutions operationally, technologically and economically feasible?

Principle 40 goes hand in hand with principle 53 relating to pragmatism.

Proposed solutions need to work in alignment with operations and must be reasonably achievable.

If your solution does not meet the three criteria of being operationally, technologically and economically feasible, it is really not a solution at all.

L & M Thought Exercise: When several solutions are operationally, technologically and economically feasible, what additional factors should be considered in the decision-making?

L & M Thought Exercise: What do the terms *operationally, technologically* and *economically feasible* mean?

SEAGOING SERVICE L & M PRINCIPLE (41): Maintain appropriate professional distance from your subordinates.

"Leaders must be close enough to relate to others, but far enough ahead to motivate them." - John Maxwell

It is difficult to be an individual's supervisor if you haven't maintained appropriate professional distance from them.

This difficulty is exacerbated if you've engaged in activities inconsistent with your organization's values and standards of conduct.

Avoid this with a passion. Do not compromise your ability to get your job done by failing to maintain appropriate professional distance from your subordinates. Once compromised, it is extremely difficult to recover.

L & M Thought Exercise: How do your seagoing service's ethics regulations address this issue?

L & M Thought Exercise: What are your seagoing services policies regarding interpersonal relations (romantic/sexual) between supervisors and subordinates? Peers in the same office?

SEAGOING SERVICE L & M PRINCIPLE (42): Expect professionalism, not perfection.

Professionalism in the seagoing services can be described as the performance of duties and responsibilities in a skillful, competent manner consistent with the expectations of the individual's organization. It involves devotion to duty, adherence with existing instructions and policies, and a willingness to continue to learn, grow, and meet expectations.

Insist on professionalism. Do not expect perfection.

You are not perfect, I'm not perfect, and neither are your subordinates.

L & M Thought Exercise: How good, is good enough?

L & M Thought Exercise: How can the Leader and Manager improve the professionalism of their staff?

L & M Thought Exercise: How can the individual improve their own professionalism?

SEAGOING SERVICE L & M PRINCIPLE (43): Stewardship.

In the context of the seagoing services, stewardship refers to Leaders and Managers responsibility to appropriately use and develop resources, including people, physical resources and infrastructure, and money. Stewardship also involves contingency planning, assuring continuity of operations, employee professional growth and development, and performance improvement.

Stewardship is important to ensuring that the necessary resources will be in place when and as needed.

L & M Thought Exercise: Describe a situation you were involved with during which stewardship was not appropriately demonstrated.

L & M Thought Exercise: Describe three ways in which the Leader and Manager can improve upon the stewardship of the people, physical resources, infrastructure, and money they are responsible for.

L & M Thought Exercise: Does everyone in a seagoing service have responsibilities in regards to stewardship? Explain your perspective.

SEAGOING SERVICE L & M PRINCIPLE (44): Thoroughness adds value; however understand the concept of diminishing marginal returns.

There exists a point beyond which each additional input of man-hours and dollar expenditures yields smaller and smaller increases in output.

Resources must be effectively distributed to assure that all missions can be met. Over devotion of scarce resources on something far in excess of what is realistically necessary can be just as harmful as under devotion of resources when viewed in the aggregate context.

L & M Thought Exercise: At what point do you accept a product or situation 'as is?'

L & M Thought Exercise: Can you cite an example of this principle that you have observed?

SEAGOING SERVICE L & M PRINCIPLE (45): Safety of life, property, and the marine environment.

Leaders and Managers, in the performance of their duties, must not needlessly endanger people, assets, and the environment.

When the mission does require putting people and resources in harm's way, risk management and mitigation efforts are required.

Does there exist a better way of achieving the same result with less risk to humans, equipment, and the environment?

L & M Thought Exercise: Which is the most important of the three (Safety of life, property, and the marine environment)?

L & M Thought Exercise: Name five means by which you can reduce risk to people.

L & M Thought Exercise: What is 'risk management?' What policies does your seagoing service have relating to 'risk management?'

SEAGOING SERVICE L & M PRINCIPLE (46): Don't be afraid to get dirty. It is part of the job.

Leaders and Managers, by the nature of their positions and responsibilities often spend a lot of time in offices. The office could be in the Pentagon, aboard a ship, or in a tent in the desert.

Getting out in the field to observe where the rubber meets the road (or the steel hits the water) creates understand of what is actually occurring. There is a significant difference in what we believe is occurring out in the field from the vantage of being in an office, than what is actually taking place.

There tends to be a lot of organizational filters which keep the guys and gals in the front office in the dark. Getting out in the field will help light a candle in the darkness and shed some light on the situation.

L & M Thought Exercise: If your duties do not normally involve field activities, what is the value of seeing what is going on in the field?

L & M Thought Exercise: For your operational specially, how often should you be getting out in the field?

SEAGOING SERVICE L & M PRINCIPLE (47): Be a positive change agent.

"If you don't like something, change it. If you can't change it, change your attitude. Don't complain."- Maya Angelou

While there are many outstanding textbooks relating to facilitating improvements, I've found the work of John P. Kotter particularly helpful. His book *"Leading Change"* is a must read.

Leaders and Managers have the inescapable responsibility to seek and promote improvement. Live up to this responsibility.

L & M Thought Exercise: What is a 'change agent?'

L & M Thought Exercise: Think about the personal and professional qualities of an individual you knew who was a change agent. What qualities did that person demonstrate?

SEAGOING SERVICE L & M PRINCIPLE (48): When a member has punitive action initiated against them, the impact extends to the member's family, the community, and your unit.

The Leader and Manager must recognize that when punitive action is administered, it impacts far more people than the person being administered the punishment.

Under the UCMJ many of you will have the authority to administer punishments involving reduction in rate and pay.

Consider the impact of your decisions. The loss of pay to a junior member doesn't necessarily mean he or she will have less money to go drinking with on Saturday night. It could mean the loss of the member's transportation, which is necessary to get him or her to and from the job, or some other unanticipated outcome. It could mean that their kids go hungry. Think about it.

L & M Thought Exercise: Presuming that you have UCMJ Article 15 Authority, what would be the potential alternatives to reducing a member in rank?

SEAGOING SERVICE L & M PRINCIPLE (49): Prevention of the death, injury, accident, and environmental harm is always preferable to a reactive response after bad outcomes have occurred. But if they have occurred, seek to prevent the next bad outcome.

When a needless death, injury, accident or environmental incident occurs it is a tragedy. When Leaders and Managers fail to take action to prevent the next incident from occurring, it is negligence.

Negligence is defined as the failure to exercise the standard of care which a reasonably careful individual would have exercised in a similar situation. Following a serious bad outcome, the reasonable and prudent person would seek to prevent the next incident from occurring (In reality, the Leader and Manager should be seeking to prevent the incidents from occurring before they actually take place through establishment of, and adherence to operational protocols and safety programs which include best practices and other control mechanisms). Following a death, injury, accident or environmental incident occurring, the Leader and Manager should consider use of a *"hot wash"* if more formal mechanisms are not used, or in conjunction with more formal mechanisms. A *"hot wash"* is a term used to describe the after-action discussions and evaluations which are performed following an incident. A *"hot wash"* session facilitates the identification of strengths, weaknesses of the response to a given event, and lessons learned. The *"hot wash"* must include the involved parties, and often serves as a 'call to action' to prevent future incidents.

L & M Thought Exercise: What are the means by which you can prevent the next bad outcome?

SEAGOING SERVICE L & M PRINCIPLE (50): Test your equipment before you need to use it.

In the seagoing services, equipment is often deployed to remote sites, potentially thousands of miles from the equipment's source, where outcomes are often dependent upon the operability of the equipment.

Pre-testing equipment in advance of an operation or shipping it to a remote site saves time, money, and effort. It could mean the difference between success and failure.

L & M Thought Exercise: Choosing a piece of equipment you are familiar with, what are the advantages and disadvantages associated with pre-testing that equipment?

L & M Thought Exercise: What is the most critical piece of equipment you rely on to perform your unit's mission? What is the outcome should it fail? Has a plan been established to address the failure of this critical piece of equipment?

SEAGOING SERVICE L & M PRINCIPLE (51): Understand the limitations of your personal protective equipment.

All personal protective equipment (PPE) has limitations.

A respirator with particulate filters won't provide the wearer with protection against organic vapors including solvents. Depending upon the material a protective glove is constructed of, it can provide either great, acceptable, or poor protection from specific chemicals used in the work environment. Different chemicals permeate different materials at different rates. Not all hearing protection affords the same protection against loud noises and hearing loss.

When you, and those you are responsible for understand the limitations of PPE, risk is better managed and the risk of acute and long-term chronic problems is reduced.

L & M Thought Exercise: Is there any personal protective equipment (PPE) which you and your subordinates should be using, but are not? What steps have you initiated to correct the problem?

L & M Thought Exercise: Do you and your team have available instructions on the use, maintenance, cleaning, and disposal of the personal protective equipment (PPE) which is used? If not, why not?

SEAGOING SERVICE L & M PRINCIPLE (52): Maintain fiscal responsibility.

As a Leader and Manager, it is critical that fiscal responsibility be maintained in three primary arenas:

1.   Adherence with procurement and financial management policies and procedures;

2.   The expenditure of money occurs throughout the period. At many seagoing service units, money expenditures are managed extremely poorly. Those in need of resources are starved for them during most of the fiscal period, and then as the period closes there is a mad effort to expend funds. This is a poor and irresponsible practice; and,

3.   Discretionary purchases need to be monitored. Even if something is permitted to be purchased under the procurement and financial policies and procedures, it may not be an appropriate purchase. Every single penny of money provided to the seagoing services represents the hard labor of an individual who paid his taxes.

L & M Thought Exercise: Is there any waste of money which you can eliminate at your unit?

L & M Thought Exercise: Have your subordinates been trained on what constitutes acceptable procurements?

SEAGOING SERVICE L & M PRINCIPLE (53): Pragmatism.

Pragmatism is generally defined to be a realistic and practical, matter-of-fact way of approaching situations and solving problems.

Pragmatism is a great approach which involves knowing when to be flexible and when to rigidly adhere to requirements. It is a balancing of capacity, needs, and outputs.

L & M Thought Exercise: Recall a situation where you or a subordinate demonstrated pragmatism. Was it a good approach to take?

L & M Thought Exercise: Under what circumstances should a different type of approach be undertaken?

SEAGOING SERVICE L & M PRINCIPLE (54): Be attentive to your external environment including personal security.

Would you walk through a high risk neighborhood alone in the middle of night while intoxicated? Some of your shipmates do this, particularly in foreign ports, with bad results.

Whether it is in the middle of the night in a foreign port, or the middle of the day on the job, you need to be attentive to your external environment.

Ships, shipyards, repair facilities, and other locations present hazards to you and your co-workers. Being attentive is something which you can do to eliminate undesired outcomes and better manage risk.

L & M Thought Exercise: Recall a situation when you or another individual were not attentive to their external environment. What was the outcome?

L & M Thought Exercise: List the five primary risks to the safety of your subordinates which exist in the external environment. How are these risks being acknowledged and minimized? Are there additional risks beyond these five? If so, have they also been acknowledged and minimized?

SEAGOING SERVICE L & M PRINCIPLE (55): When overseas, respect the laws and culture of the host nation.

As members of the seagoing services of the United States, our conduct is closely scrutinized by citizens and governments of the nations we visit. The conduct of one individual has the capacity to destroy or seriously damage positive international relations which have been slowly built up through time.

If you do not obey the laws of the nation you are visiting and become incarcerated, most likely you will not be afforded the legal protections afforded within the United States. You may not be afforded legal counsel or bail. Prisons overseas, while slowly improving as a result of efforts by the United Nations, are unpleasant at best and hell-holes at worse where brutality and violence are the norm.

Be the 'Good American' while overseas, not the 'Ugly American.'

L & M Thought Exercise: Recall a situation when either you or someone else did not respect the laws and culture of the host nation. What was the outcome?

L & M Thought Exercise: If you break the laws of another nation, could you potentially face punitive action under the UCMJ? Please explain.

SEAGOING SERVICE L & M PRINCIPLE (56): Innovate.

"Even if you are on the right track, you will get run over if you just sit there." – Will Rogers

Innovation involves incremental to radical improvements in processes, products, services, and strategy. Innovation means doing things better today than they were done in the past, and making sure they're done better tomorrow than today.

Innovation promotes mission accomplishment. It is critical.

L & M Thought Exercise: Think about the most three significant innovations in your field or specialty in the last ten years. What were they?

L & M Thought Exercise: Think about a specific situation in your working environment which could be improved upon. What innovations can be implemented?

SEAGOING SERVICE L & M PRINCIPLE (57): Shooting-from-the-hip is best left to the Westerns.

'Someone who shoots from the hip' is an idiom which refers to a person who makes decisions, or talks very directly or insensitively without thinking beforehand. More succinctly, it means to act or speak on a matter without forethought or foresight.

Rash decisions and communication are a great way to achieve poor, sub-optimized results. You have the capacity to do better than this.

If you learn the difference between 'shooting-from-the-hip' and expeditious decision-making, you and your team will benefit significantly.

L & M Thought Exercise: Think about the last time you saw someone shoot from the hip. Were the desired outcomes achieved?

L & M Thought Exercise: Expeditious decision making involves what thought processes and actions by the decision maker?

SEAGOING SERVICE L & M PRINCIPLE (58): Understand the end goal, the desired outcome.

Failure to understand the end goal is like driving in your car without having any idea of your final destination. When you are driving without a destination, you are expending resources (fuel and time), there is the sense of progress, and the scenery (external environment) changes. But are you really getting to where you need to go?

Understanding the end goal requires vision, a set of goals to take you from where you are presently located to a place further down the highway, and clear objectives.

Do you and your team understand the end goals, the desired outcomes?

L & M Thought Exercise: Recall a situation when the end goal, the desired outcome, wasn't clearly understood. Were results optimized?

L & M Thought Exercise: How can Leaders and Managers improve communication and understanding in this area?

SEAGOING SERVICE L & M PRINCIPLE (59): Wear your safety equipment.

This goes hand-in-hand with an understanding of the limitations of personal protective equipment. Most likely a trained and qualified safety professional has made an assessment that the safety equipment you need to wear is necessary to protect you and your staff from short and long term health problems.

During periods of exposure, often the person being exposed does not feel any ill effects. They are experienced later on, at which point it is too late.

Read and understand the Material Safety Data Sheet (MSDS) for each chemical you could be potentially exposed to. It could save your life.

L & M Thought Exercise: In your work environment, what are the potential bad outcomes if you do not wear your safety equipment (Personal Protective Equipment)?

L & M Thought Exercise: Where are the MSDSs located at your worksite?

L & M Thought Exercise: Examine three MSDSs. What features do all three have in common?

SEAGOING SERVICE L & M PRINCIPLE (60): Limit occupational exposures.

Occupational exposures are the exposures we get to chemicals, noise, vibration, and energy in our work environment.

The Occupational Safety and Health Administration (OSHA) has established Permissible Exposure Limits (PELs). The American Council of Governmental Industrial Hygienists (ACGIH) have also established protective standards, known as 'Threshold Limit Values (TLVs) which are generally lower than the federal limits. Just because an occupational exposure is lower than a PEL or TLV doesn't mean with absolute certainty that the person being exposed to is safe. Human physiology results in some people demonstrating greater sensitivity to exposure than others, various exposures can create a synergistic effect (1 + 1 = 3), and ongoing research has consistently established lower and lower safety limits. Seek to limit exposures to levels below the most conservative of the PEL and TLV, and keep the exposures *"As Low as Reasonably Achievable* (ALARA)."

L & M Thought Exercise: Think about your most recent occupational exposures. How could they have been prevented or significantly reduced?

SEAGOING SERVICE L & M PRINCIPLE (61): Measurement and quantification in terms of distance, volume, mass, thickness, time and other metrics results in understanding, precision and ability to duplicate.

The use of metrics and quantification promotes better decision making. It is just that simple.

If you were the Engineering Officer aboard your ship, which report would you prefer from your subordinate?

"Commander, we are running low on diesel fuel."

"Commander, we are running low on diesel fuel. We have 98,000 gallons remaining. At our present burn rate we have an eleven day supply. At maximum potential burn rate, we have an eight day supply. It will take five days to coordinate refueling in our present operating area. I recommend that given..."

L & M Thought Exercise: Think about what you can quantify on your job that would promote better decision making or evaluation.

L & M Thought Exercise: Is it possible to over quantify on your job? Describe a situation of over quantification if you believe it is possible to over quantify.

SEAGOING SERVICE L & M PRINCIPLE (62): Use of appropriate terminology conveys understanding, experience and professionalism.

Technical terminology is the specialized vocabulary of a profession or specialty. Within a specialty, the terminology has a specific meaning which may be different than what the term means outside of the specialty. Jargon, or slang, within a specialty is similar, but more informal.

The use of appropriate terminology promotes communication between individuals without regard to location. It denotes experience and professionalism. For your specific specialty, it is absolutely critical.

L & M Thought Exercise: Recall a situation where the lack of using appropriate terminology created confusion or misunderstanding of a situation.

L & M Thought Exercise: List the ten most important terms or phrases specific to your specialty. For example, in the legal profession, the terms might include *'jurisprudence'* and *'tort.'* Then list an additional ten.

SEAGOING SERVICE L & M PRINCIPLE (63): Recognize trends.

The ability to recognize trends and understand the implications of the trend is critical for making future decisions relating to resources such as people, money, and physical assets. Proper trend analysis gives substance to your 'gut' instinct of where things are going.

Trend estimation can range from extremely simple to more complex use of statistical techniques to aid interpretation of data. When a time series of measurements of a process exists, the application of trend estimation can be used to make and justify statements about trends in the data and facilitate the construction of a model which is independent of anything known about the physics of a process of an incompletely understood physical system. It is useful to determine if measurements demonstrate an increasing or decreasing trend that are statistically distinguished from random variance.

L & M Thought Exercise: Once you've recognized a trend, what's next?

L & M Thought Exercise: Many computers in the seagoing services come equipped with graphing and statistical software. If your computer is so equipped, do you understand how to use the software? If no, why not?

SEAGOING SERVICE L & M PRINCIPLE (64): When things go wrong, keep it all in perspective.

"Check up each week on the progress you are making. Ask yourself what mistakes you have made, what improvement, what lessons you have learned for the future." - Dale Carnegie

I recall having an Executive Officer who kept it all in perspective by saying that as long as no one had died or been injured we are doing okay. While he was not exactly correct, since even if there were no deaths or injuries the organization could be failing, it was a good commentary on proper perspective.

Regardless of whether you are in a seagoing service or in the private sector, there are going to be times when things go wrong despite proper preparation and due diligence.

Sometimes you have to just roll with the punches. Learn and move on.

L & M Thought Exercise: When things go **right**, do you also need to keep it all in perspective?

SEAGOING SERVICE L & M PRINCIPLE (65): Create effective and appropriate efficiencies.

Streamlining processes and procedures can create effective and appropriate efficiencies provided there is complete understanding of the implications of your changes.

Remember that the process was initially created by someone who developed it to eliminate a specific problem, or to meet an operational need, or to adhere to a mandated policy.

Once the process or procedure is streamlined, document it so that in the future it will be adhered to. Over time, undocumented processes and procedures degrade and gradually return to a state of inefficiency. Formalizing the improvements will ensure that they 'stick' and the efforts of the team are not wasted.

L & M Thought Exercise: Think about the one thing you could improve in your work environment which would create additional efficiency, effectiveness, or quality. Can the barriers preventing the improvement be removed?

SEAGOING SERVICE L & M PRINCIPLE (66): We have multiple customers.

"As a global company, our future growth and success requires that we constantly look at ways to improve our ability to serve customers worldwide." - Steve Ballmer

A customer is someone we provide a service or product to. Our customers include our Command Cadre, peers, subordinates, and external parties we are involved with.

Remember, one of our customers is 'John Q. Public' – the man and woman on the street who by dint of their labor provides the funding and support for your seagoing service. Do not forget that as members of the military we have inescapable responsibilities to our customers, most notably the public.

L & M Thought Exercise: Who are your specific customers? What are their specific needs?

L & M Thought Exercise: What is a supplier? Who are your suppliers?

L & M Thought Exercise: What are our inescapable responsibilities the public?

SEAGOING SERVICE L & M PRINCIPLE (67): Not all issues are of equal importance.

As a Leader and Manager on any given day you have multiple priorities and deliverables. A few minutes spent organizing the next day's issues in terms of importance and due dates will help you psychologically prepare for the next day, promote focus and effectiveness.

Remember – the critical issues and show stoppers rarely, if ever, go away until they have been satisfactorily and professionally addressed.

L & M Thought Exercise: What is the most important issue you need to address today? This week?

L & M Thought Exercise: What are the criteria which you use to prioritize your work? Is there a better way to prioritize?

SEAGOING SERVICE L & M PRINCIPLE (68): What is the root cause?

When there exists non-conformance with requirements, the Leader and Manager needs to determine the root cause for the non-conformity.

Determining the root cause of non-conformity can be achieved by conducting a root cause analysis, which is a structured evaluation method that identifies the root causes for an undesired outcome. The National Aeronautics and Space Administration's (NASAs) Office of Safety & Mission Assurance considers a root cause to be *"One of multiple factors (events, conditions or organizational factors) that contributed to or created the proximate cause and subsequent undesired outcome and, if eliminated, or modified would have prevented the undesired outcome. Typically multiple root causes contribute to an undesired outcome."*

The six steps in root cause analysis, according to NASA, are:
(1) Defining the undesired outcome;
(2) Gathering data, including all potential causes;
(3) Creating an event and causal factor tree;
(4) Continuing to ask "why" to identify root causes;
(5) Checking the logic which was used, and eliminating items that are not causes; and
(6) Generating solutions that address both proximate causes and root causes.

Additional information on this very useful tool is widely available on the internet and professional publications.

L & M Thought Exercise: Describe a situation in which knowledge of the root cause of an incident or situation promoted better decision making.

SEAGOING SERVICE L & M PRINCIPLE (69): Carry up to date business cards.

Leaders and Managers meet a lot of other people who can assist them in resolving issues.

An up to date business card facilitates future communication.

Avoid humor, offensive material and anything else which might be misunderstood on your card.

Keep it simple: name, organization, business address, e-mail, telephone, facsimile, web site address and corporate logo.

If you want anything else on the card, think about how it may add or detract value.

L & M Thought Exercise: Do you have sufficient interaction with others, particularly people external to your organization, to merit having business cards? If so, do you have them?

L & M Thought Exercise: If you have business cards, are they current? Are the cards which you present to others new and crisp, or are they battered, worn and bent from being crushed in a wallet? Remember, your card is an extension of your professionalism.

SEAGOING SERVICE L & M PRINCIPLE (70): Recognize the power centers.

Recognizing the power centers, that is who the decision makers are, allows the Leader and Manager to focus their efforts on that specific individual or that specific office.

It eliminates non-productive efforts, provides focus, and eliminates the "shot-gun" effect.

L & M Thought Exercise: Who are the 'power centers' you routinely work with?

L & M Thought Exercise: What are the advantages associated with recognizing power centers?

SEAGOING SERVICE L & M PRINCIPLE (71): A picture can speak a thousand words.

Photographs and recordings are a great tool for documenting and quantifying (See Principles 31 and 61). They are a means by which Leaders and Managers can communicate findings, observations, concerns, and events.

Ideally, photographs should capture events from multiple angles (overhead, and view from four points), show orientation and detail, and have a means to determine scale. Information relating to date, time, and location the photograph was taken and subject information is also important.

L & M Thought Exercise: What are your service's policies relating to taking photographs of operations?

L & M Thought Exercise: Under what circumstances should you and your team not document events and issues with photography?

SEAGOING SERVICE L & M PRINCIPLE (72): Don't be too accepting of industry hospitality.

"With integrity you have nothing to fear, since you have nothing to hide. With integrity you will do the right thing, so you will have no guilt. With fear and guilt removed you are free to be and do your best." - Zig Ziglar

Very few things in life are free. Industry hospitality comes with a cost in terms of expectations and consideration.

Always stay within your service's ethical standards when it comes to accepting industry hospitality. Exceed them (take a more conservative approach, stricter adherence) when possible.

L & M Thought Exercise: What does your service's ethics policies say about accepting industry hospitality?

L & M Thought Exercise: What interactions with industry and external parties take place in your immediate work environment which could create or raise the appearance of ethical conflicts and improprieties? How are these issues addressed?

SEAGOING SERVICE L & M PRINCIPLE (73): From most effective to least effective, the hierarchy of health and safety controls are elimination or substitution, engineering controls, warnings, training & procedures, other administrative controls, and personal protective equipment.

Why should you care about the hierarchy of health and safety controls? Outstanding question.

As a Leader and Manager, you have the responsibility for providing your subordinates with a safe working environment. The hierarchy of controls provides you with a broadly based roadmap on how to proceed from where you are at to a safer environment.

The use of personal protective equipment is the least preferred course of action (other than totally ignoring safety hazards) because it relies on the human element which includes the proper wearing of the equipment – a situation which doesn't always occur, except in the most heavily supervised work environments.

L & M Thought Exercise: What are your responsibilities in regards to providing a safe working environment for your subordinates?

L & M Thought Exercise: Who is responsible for safety within your seagoing service?

Accident investigators estimate that as much as 80-85% of accidents are related to human factors such as fatigue.

Research by the federal government has identified four primary fatigue factors: the circadian rhythm effects, sleep deprivation, cumulative fatigue effects, and "time-on-task" fatigue.

Circadian rhythm effects describe the tendency for people to experience a normal cycle in attentiveness and sleepiness through the 24-hour day. Those with a conventional sleep pattern of sleeping for seven or eight hours at night experience periods of maximum fatigue in the early hours of the morning and a lesser period in the early afternoon. During the low points of this cycle, one experiences reduced attentiveness. The influence of the day-night cycle is never fully displaced, and the performance of night shift workers usually suffers.

Sleep deprivation involves individuals who fail to have an adequate period of sleep, usually eight hours in a twenty-four hour period, or who have been awake longer than sixteen hours.

Cumulative fatigue results from the accumulation of sleep-deprived days and broken sleeping patterns. It can often take individuals two to three days to recover from a sleep deficit.

"Time-on-task" fatigue describes fatigue that is accumulated during the working period. Performance declines the longer a person is engaged in a task, gradually during the first few hours and more sharply as the duration of the shift continues.

As Leaders and Managers we need to be attentive to this issue. Often the most dangerous period of time occurs after the end of the shift, when the individual, no longer pumped up by ongoing events, falls asleep behind the wheel of his or her car. The results of excessive fatigue are often deadly, and with proper management of resources, can under most circumstances be avoided.

L & M Thought Exercise: Have you ever endangered yourself or others because of excessive fatigue? If so, how are you preventing recurrence of the dangerous situation?

SEAGOING SERVICE L & M PRINCIPLE (75): An operation takes as long as it needs to take.

When planning operations, the Leader and Manager must recognize that delays may and do occur due to events which were reasonably unpredictable. Don't plan for the optimal duration of an event, plan for the most realistic time and then add a cushion.

When we rush, we lose quality, attention to detail, and with many operations we create unnecessary risk of undesired outcomes – death, injury, loss of infrastructure, environmental harm et cetera.

When the undesired outcomes occur (and they will if we rush an operation), the end result is far worse than what would have resulted if we had allowed sufficient time.

Slow down, and get to the finish line quicker.

L & M Thought Exercise: Do you have enough time to do your job professionally and safely? If no, what are you doing to correct the situation?

L & M Thought Exercise: Take the time to undertake a series of measurements on your most critical operation and perform a statistical analysis. Based on this analysis, should your organization be allowing more time for the critical operation?

SEAGOING SERVICE L & M PRINCIPLE (76): Intoxication increases personal risk.

Intoxication increases personal risk of accidents, injuries, death, and other undesired outcomes impacting your career and interpersonal relations. It takes less than you think. Though every individual has a different tolerance to alcohol, on average the following physiological reactions take place: Approximate BAC = 0.03 to 0.12%: Increased self-confidence and courage which means that when you should potentially be fearful of something such as risky behavior you might not be; shortened attention span; flushed appearance; inhibited judgment; and impaired fine muscle coordination, such as writing or signing their name. Approximate BAC = 0.09 to 0.25%: Sedation; impaired memory and comprehension; delayed reactions; balance difficulty; walking is not stable; and blurred vision; other senses may be impaired. Approximate BAC = 0.18 to 0.30%: Profound confusion; dizziness and staggering occur; speech is impaired; impaired senses; analgesia and vomiting. Approximate BAC = 0.25 to 0.40%: Severe ataxia; lapses in and out of consciousness; unconsciousness; amnesia; vomiting; dangerous respiratory depression; and potential death. Approximate BAC = 0.35 to 0.50%: unconsciousness; severely depressed reflexes and respiratory depression; and death usually occurs at levels in this range.

Each of the seagoing service's have policies relating to intoxication. These were established for good reason. A significant portion of Court Martial's, Article 15s, and discharges for cause involve alcohol and/or drugs.

L & M Thought Exercise: Think about the bad outcomes associated with intoxication that you've seen. Were the outcomes worth it to the person they occurred to?

SEAGOING SERVICE L & M PRINCIPLE (77):
The Pocket Memo Book.

Leaders and Managers, by the very nature of their positions, are busy people. They must respond to issues which 'pop' up at every turn. A walk from one end of the ship to another can result in three of four taskings and issues which must be resolved.

Few people are sufficiently focused and organized enough to remember all of these issues without writing them down.

A pocket memo book helps you to address these issues without having them 'slip through the cracks.' This small practice helps you to be more organized and responsive to others.

L & M Thought Exercise: Do you maintain a pocket memo book? If not, do you have a mechanism for the immediate capture of unexpected tasking?

SEAGOING SERVICE L & M PRINCIPLE (78):
Walk-the-Talk.

Walk-the-talk means to do as you say you would do, to be consistent with your words and actions. It is an important component of professional and personal credibility.

Those who walk-the-talk are individuals who meet their commitments and promises.

L & M Thought Exercise: Of the three individuals you work most closely with, do they all *'walk-the-talk?'* How does their adherence with this principle impact the level of trust and confidence you have in these individuals?

L & M Thought Exercise: What are the implications of failing to walk-the-talk?

SEAGOING SERVICE L & M PRINCIPLE (79):
RHIR, not RHIP.

Rank has its responsibilities (RHIR) not privileges.

While realistically there are a few privileges associated with rank and position, you are still required to adhere to your organization's policies, procedures, core values and the Uniform Code of Military Justice.

Your position does not afford you the right to ignore the standards that each member of your organization is required to adhere to. It imposes additional responsibilities because of the authority which has been afforded to you as a result of your rank or position.

L & M Thought Exercise: Describe some legitimate privileges associated with rank and position. Compare these with the increased responsibilities associated with the rank and position.

SEAGOING SERVICE L & M PRINCIPLE (80):
Be sensitive to pay grade and positional power differences.

Many individuals make requests of peers for assistance with personal matters - 'favors' large and small that our peers are generally pleased to provide support on, since they know that we will eventually return the 'favor.'

In superior – subordinate relationships you need to be sensitized to the issue of pay grade and positional power differences. Because of pay grade and positional power issues, often the subordinate feels unduly pressured into performing the 'favors' and the 'quid-pro-quo' relationship that one had with their peers no longer exists.

This is a potential recipe for disaster, and may result in potentially supported misconduct charges be leveled against the superior.

L & M Thought Exercise: Do you have subordinates take care of your personal business, which rarely, if ever, is appropriate?

L & M Thought Exercise: Have you ever felt imposed upon when a superior required you to perform a task which was within the realm of 'personal business?'

SEAGOING SERVICE L & M PRINCIPLE (81):
Contributions.

In the seagoing services we often provide contributions for retirement gifts, births, deaths and significant life events. At some units, the coffee mess is funded through voluntary contributions.

If you are grabbing cups of Joe for yourself or guests from a self-funded coffee mess, are you making fair contribution? You should be. Depending on your rank or the positional power you exert over people, they may not confront you on the issue. They do however care!

Are you contributing to the solicitations for donations to pay for the small gifts for births, deaths et cetera? Though voluntary, it is appropriate that you send the message of caring, concern, and compassion. Your contribution helps to send the appropriate message.

L & M Thought Exercise: Are your contribution's fair?

L & M Thought Exercise: Are you proactive in your contributions, or do you require people to track you down and make a personal solicitation?

SEAGOING SERVICE L & M PRINCIPLE (82):
Acknowledging human milestones.

Members of the team have lives outside of the workplace. Employer acknowledgement of significant events, both positive and negative, such as marriage, the birth of a child, and death is important to people.

This is particularly important for people who may not have an adequate support network away from the work environment, or for those geographically separated from their family due to deployment.

L & M Thought Exercise: Make a comprehensive list of significant personal events which are recognized in the work place.

L & M Thought Exercise: What are the human milestones in your life which you would like to have recognized? Have you acknowledged others when they met a similar milestone?

SEAGOING SERVICE L & M PRINCIPLE (83):
The three dimensional employee.

By and large, your subordinates are smart and talented people. They have skill sets which many of them derive great joy from using, though they may not be normally used or recognized in the work place.

By viewing your subordinates three dimensionally as individuals with a wealth of talent, you can often tap into their unique skills to meet mission.

For example, recently my unit's Executive Officer retired from our seagoing service. Because the ceremony was held overseas, many people who desired to attend were simply unable to do so – including the retiree's parents. One of the members of our unit, who is working after hours on a Master's of Fine Arts degree in Film Directing, viewed the ceremony as an opportunity to convey respect towards the retiree and as a chance to use his acquired film directing skills. Result: The development of a fifteen minute 'short' showing the highlights of the retirement ceremony which was sent to all invitees.

L & M Thought Exercise: Describe a unique talent which you possess. Are others in your work environment aware of this talent?

L & M Thought Exercise: Describe a unique talent which one of your subordinates possesses.

SEAGOING SERVICE L & M PRINCIPLE (84):
Moving beyond disappointment.

"One's best success comes after their greatest disappointments." - Henry Ward Beecher

"The most vital test of a man's character is not how he behaves after success, but how he sustains defeat."- Raymond Moley

The Leader and Manager at various points in their career will experience professional disappointment, as well as serving as the supervisor of individuals who also experience disappointment.

Though difficult, we each need to move on following the disappointment. Seeking understanding as to why the we did not obtain what we sought, and learning from the experience is one of the best things an individual can do when we get 'bad news.'

L & M Thought Exercise: What did you learn from your last significant professional disappointment?

L & M Thought Exercise: Describe a method which might be useful to an individual seeking to move beyond a professional disappointment.

SEAGOING SERVICE L & M PRINCIPLE (85):
Recognition.

Recognition, either formal or informal, is a public acknowledgement of actions above and beyond which is expected, and is a mechanism to motivate, foster morale and establish esprit de corps.

Recognition is also important because it communicates to members of the team that their work is valued and appreciated; provides a feeling of ownership and affiliation in their place of work; improves loyalty; builds a supportive work environment ; and when genuinely provided contributes to employee retention.

The Leader and Manager is well served by having knowledge of the policies within their seagoing service relating to personal and team recognition.

L & M Thought Exercise: What policy or instruction within your organization addresses informal and formal recognition?

L & M Thought Exercise: Are you over-recognizing your staff for routine accomplishments?; or are you under-recognizing these individuals for significant accomplishments? What measure or criteria are you using when deciding whether to initiate recognition?

SEAGOING SERVICE L & M PRINCIPLE (86):
Operating within the legal construct.

Every Leader and Manager must perform their duties within the legal construct, which establishes requirements and prohibitions on activities which must be undertaken on behalf of the organization.

The policies which your seagoing service have in place traditionally cite the law(s) and federal regulations which either mandate action or constrain you. Often, these policies summarize the law and federal regulations as they apply to professionals such as yourself.

Should you have a need to find the specific verbiage within a law or federal regulation, your legal services officer is available to provide assistance. As an alternative, the texts of laws and federal regulations may be found on the Government Printing Office *"GPO Access"* web site at http://www.gpoaccess.gov.

L & M Thought Exercise: What information cannot be released under the Privacy Act (5 U.S.C. §522a)?

L & M Thought Exercise: What laws govern the operations your unit routinely performs?

"The most powerful and predictable people-builders are praise and encouragement." - Brian Tracy

Motivation is intrinsic or extrinsic goal-oriented behavior which derives from human desire to, either minimize pain and maximize pleasure; or to meet specific needs or goals; or to benefit others or themselves. Intrinsic motivation is related to the inherent rewards associated with the task, whereas, extrinsic motivation is external to the performer and may involve rewards such as promotion, time off, formal or informal recognition, or avoidance of punishment.

As a Leader and Manager, by this time in your development you most likely have been exposed to Abraham Maslow's Hierarchy of Needs and Frederick Herzberg's Two-Factor Theory. If not, a search of the internet will provide you with numerous sites which will improve your understanding of these two accepted motivation theories.

Professor Steven Reiss, author of the widely acclaimed and worthwhile book *"Who am I: The 16 basic desires that motivate our actions and define our personalities"* and professional articles, has postulated a theory in which sixteen desires motivate most human behavior.

As a Leader and Manager, it is worth being cognizant of this theory. The sixteen desires postulated by Professor Reiss are:

- The need for approval (acceptance);
- The need to think (curiosity);
- The need for food (eating);
- The need to raise children (family);
- Loyalty to the traditional values of one's ethnic group (honor);
- The need for social justice (idealism);
- The need for individuality (independence);
- The need for exercise (physical activity);
- The need for influence (power);
- The need for sex (romance);
- The need to collect (saving);
- The need for friends and peer relationships (social contacts);
- The need for social standing/importance (status); and,
- The need to be safe (tranquility)

An understanding of what motivates groups and individuals will allow you, the Leader and Manager, to improve team and individual performance.

L & M Thought Exercise: Think of an example on how an unmotivated individual was 'turned around' to become a better performer. How did the Leader and Manager accomplish this?

SEAGOING SERVICE L & M PRINCIPLE (88):
Understanding of economies and diseconomies of scale.

Economies of scale is an economic concept and refers to reductions in unit cost as the size of an operation increases. Diseconomies of scale is just the opposite, unit costs increase as the scale increases. In the seagoing services, a unit cost involves time, money, and the use of assets. In a nutshell, this principle involves maximizing output per unit of input - 'Rightsizing' of operations. For example, let us presume that your vessel has access to a van which is used for ship's business. The van holds ten people; one driver and nine others. All other things being equal, would you send a team of ten to the range (an hour's drive from the ship) for a day of small arms practice and qualification, or would you send eleven? Optimally, you would want to qualify as many people as possible, however the concept of diseconomies of scale indicates that it would be best to send the team of ten rather than eleven because the one additional member would triple the number of round trips the van would need to make to and from the range. A team of ten involves the ten getting in the van, driving to the range where the vehicle is parked, and then returning to the ship at the end of the day. A team of eleven involves driving most of the team to the range, a return drive back to the ship to pick up the remaining member, driving the remaining member to the range and then having to repeat the entire process on the return trip.

L & M Thought Exercise: Describe another situation involving economies and diseconomies of scale which is relevant to your work situation.

SEAGOING SERVICE L & M PRINCIPLE (89):
Performance measurements.

Performance measurements allow the Leader and Manager to clearly determine their unit's performance compared to desired benchmarks.

Renowned researcher, Robert D. Behn, author of *Why measure performance? Different Purposes Require Different Measures*, believes that performance measures are necessary to evaluate outcomes and cost-effectiveness, to control human work related activities, to budget deployment of financial resources, to motivate performance, to celebrate reaching of milestones and specific goals, to promote the organization and its staff, to learn the reasons behind good or poor performance, and to improve.

Without measurement, there exists no specific, focused knowledge on where and what to improve. The use of 'gut' feelings and 'instinct' sometimes has merit, however the quantification process results in decision optimization.

L & M Thought Exercise: Are those items that you are measuring focusing attention on critical success factors? If no, why not?

L & M Thought Exercise: Are performance measures defined in terms of responsibility, measurement unit, frequency of data collection, quality of data, targets, and thresholds? If no, why not?

SEAGOING SERVICE L & M PRINCIPLE (90):
Advertising of services.

Leaders and Managers advertise the services of the unit within their area of responsibility. If the rest of the team is not aware of what your organizational component brings to the table, then optimization of resource utilization isn't achieved. To paraphrase an old Army recruiting campaign slogan, *'the team isn't all that it can be!'*

Your communication can facilitate the introduction of new services, reinforce the existing knowledge of what your component has to offer, eliminate misconceptions and increase utilization of your team's services.

L & M Thought Exercise: Are you absolutely certain that your customers have a complete and accurate understanding of the services provided by your organizational entity? If not, what can be done to improve the situation?

L & M Thought Exercise: What are the mechanisms your organization can use to increase understanding of what your organizational component brings to the table?

A corollary of maximizing desired outcomes is the concept of minimization of least desired outcomes. Sometimes it is more important to minimize the potential for least desired results than to optimize outcomes.

This involves an understanding of consequences. For example, for a ship underway in a storm, the maximized desired outcome may be meeting a previously arranged operational schedule. Operational schedules are important – without an operational need, there exists little reason to get underway and deploy expensive human and physical resources.

Using the principle of minimization of least desired outcomes, the Commanding Officer may be forced to minimize the least desired outcome, the loss of, or severe damage to their ship rather than to achieve the preferred maximization of desired outcomes.

This is a principle worth remembering.

L & M Thought Exercise: Cite an example of this principle involving your unit.

L & M Thought Exercise: Is it more important to maximize desired outcomes or minimize least desired outcomes?

SEAGOING SERVICE L & M PRINCIPLE (92):
But Sir...

There is a mutual responsibility which exists in the superior – subordinate relationship. As a superior, you have the responsibility to listen to and consider the input of your subordinates. None of us are infallible, and none of us makes perfect decisions 100% of the time given our own inherent biases and filters. As a subordinate, you have the professional responsibility to appropriately express your concerns when it appears that decision-making is off track. Sometimes, your responsibility requires you to express your 'But Sir...' or 'But Ma'am...'. This is a difficult thing to do, particularly when your superior is known for having a strong personality.

Some thoughts:

(a)   Pick and choose your issues wisely. If, once in a blue moon, you express concern on a major issue, most likely you will be listened to very closely. If you are known for expressing your concerns on every issue from the most minor to more pressing, your concerns most likely won't receive the same consideration; (b) Communicate clearly, factually, and tactfully while acknowledging that the responsibility and final decision rests with your superior; and, (c) Know when to stop pressing your concerns. Many superiors appreciate hearing concerns. But once they've heard them, evaluated them, and rendered a decision in the matter, you need to stop pressing the issue and trust upon the experience and judgment of your superior.

L & M Thought Exercise: Cite an example of this principle which you observed.

SEAGOING SERVICE L & M PRINCIPLE (93):
Don't throw gasoline on the fire.

The phrase *"throwing gasoline on the fire"* refers to taking action which makes an undesired or extremely bad situation worse.

As a Leader and Manager, you have the capacity to make existing and future situations better or worse.

To the extent reasonably feasible, and in a manner consistent with good order and discipline, and other seagoing service policies, seek to improve the situation. Avoid throwing gasoline on the fire.

L & M Thought Exercise: Describe a situation you are familiar with in which an individual figuratively *"threw gasoline on the fire."*

L & M Thought Exercise: Using the above situation, what were some viable alternatives to the *"throwing gasoline on the fire"* which could have been potentially used?

SEAGOING SERVICE L & M PRINCIPLE (94):
Strike while the iron is hot.

Term *'strike while the iron is* hot', according to the 1894 *Dictionary of Phrase and Fable* which was written by E. Cobham Brewer, is a metaphor relating to a blacksmith molding a piece of red-hot iron into place.

For the Leader and Manager, this means taking advantage of timing and external events to meet mission; do what you do at the right time to optimize results.

L & M Thought Exercise: Describe a situation you are familiar with in which an individual figuratively *"struck while the iron was hot."*

L & M Thought Exercise: Think of the events which occur during a fiscal year such as budgeting, planning, training et cetera. Are some of these activities best initiated during specific time frames? Why or why not?

SEAGOING SERVICE L & M PRINCIPLE (95):
There is a time and place for yelling at people.

There is a time and place for yelling at people. The first situation is to warn people of a dangerous situation or a call to action. For example, yelling the words **_"Run, Stop, Move, Watch Out, Fire"_** to communicate that a serious danger exists and a very specific action is required immediately to prevent harm, injury or death is appropriate. The second situation might be upon observing conduct or action so inconsistent with organizational policies, practices, or core values that a very public, instantaneous message needs to be sent regarding the unacceptability of the conduct. For example, while walking through a space where a group of sailors is gathered, and you overhear a racist joke being told (Remember, just because you've yelled at someone, doesn't mean you can't pursue additional action). Yelling at people beyond these situations (which in reality occur relatively infrequent) is often counter-productive in that long term resentment is created, motivation is decreased, the possibility of an escalation of aggression increases. The list could go on and on, but you get the point. Yelling at people is a tool which needs to be utilized far less than most people realize. As a Leader and Manager, you've got better tools available to you.

L & M Thought Exercise: When was the last time you yelled at someone in your seagoing service? In retrospect, was the use of this tool necessary? What were the benefits and detriments associated with this approach?

L & M Thought Exercise: What constitutes prohibited verbal abuse in your seagoing service? What policy within your seagoing service addresses this issue?

SEAGOING SERVICE L & M PRINCIPLE (96):
Consistency without rigidity.

"We want a system that will improve consistency and steadiness in the quality of government."
- Ferdinand Mount

Consistency of good leadership and management provides external parties including superiors, peers, subordinates, and customers with an expectational framework upon which to respond to and initiate action. If Leaders and Managers adhere to the principles within this text there is an expectation that the Leader and Manager will do so in the future. This creates an environment for positive change and growth amongst others. For example, if you consistently treat people with respect (Principle 19) and adhere to your organization's core values (Principle 15), other individual's conformance with these principles will steadily improve over time.

The application of leadership and management practices though consistent, must not be rigid. What does this mean? This question is best addressed through an illustrative example. As a Leader and Manager, you should consistently respect people and insist upon the same conduct amongst others, however your application of this principle recognizes that there is no 'one size' fits all solution. Individual expectations in regards to respect differ, and an approach which works with some or most, certainly may not be suitable for all.

L & M Thought Exercise: Is your leadership and management consistent without being rigid? What can you do to improve in this area?

SEAGOING SERVICE L & M PRINCIPLE (97):
Vision Statement.

Vision statements were once in fashion in the realm of leadership and management. They now appear to be on the wane. Nonetheless, you should be an advocate of vision statements and have one posted in your common spaces (larger than 8 ½ x 11) as well as in your office or work space. The lettering needs to be in big, bold type readable from a distance. Why? Because a well crafted vision statement serves as a tool to keep your team on track and focused. In the seagoing services there are many thousands of pages of policies which have been promulgated. These policies don't cover every single decision which you and your subordinates need to make. In the absence of policy, a vision statement guides.

> USCG XXXXXXX
> INVESTIGATION DEPARTMENT VISION
>
> We will promote the safety of life, property, and the marine environment through appropriate, balanced professional inquiry into marine casualties, and the activities of mariners and the marine industry. In the conduct of such inquiry we will respect the rights of mariners, the marine industry and the public. We will seek to comply with the fundamental principles of the Coast Guard's Investigations and Enforcement Program and policies related to the program.

L & M Thought Exercise: Does your organizational entity have a vision statement which is clear, concise and understood by the team? If not, why not?

SEAGOING SERVICE L & M PRINCIPLE (98):
Morale matters.

Morale can be defined as an individual or collective state of emotional spirits exhibited by cheerfulness, discipline, and willingness to perform assigned tasks. It is linked to emotional well-being and its absence results in reduced performance and diminished feelings of esprit de corps.

Many Leaders and Managers believe that morale can be created through 'morale events.' They are incorrect. The core of morale at a unit is establishing a work environment in which people are fully provided the tools and training they need to meet tasking, the work place climate is not hostile (people are treated with respect and dignity), and there exists some flexibility to allow people to meet their needs and the things which really matter to them. For example, allowing a working mother to shift her schedule by thirty minutes so that her child can attend day care is worth more than a hundred morale events to her. Allowing one of your hard charging enlisted members to come in early so that they can attend after-hours courses at the local university serves to build their morale. In addition to establishing an acceptable org climate, Leaders and Managers need to support morale activities and events while balancing mission needs with the 'down time' that these morale events create. Morale events are an important tool – provided that there already exists the environment in which people are fully provided the tools and training they need to meet tasking, the work place climate is not hostile (people are treated with respect and dignity), and there exists some flexibility to allow people to meet their needs and the things which really matter to them.

L & M Thought Exercise: Cite an example of this principle involving your unit.

SEAGOING SERVICE L & M PRINCIPLE (99):
Fix the problem not the symptom.

Symptoms are manifestations of deeper organizational problems. For example, let us say that your seagoing service has a very low re-enlistment rate among the personnel completing their initial obligated service period (their 'first hitch'). While in reality this may or may not be true, but for illustrative purposes we will presume it to be the case.

The low retention rate is not the problem. It is a symptom of less obvious issues which may include internal leadership and management problems, failure to provide acceptable wage and benefit packages compared to private sector employers, employee burnout from unsustainable high operation tempos, or other issues.

To get to the root of a problem, ask the question 'Why is the outcome occurring?" When you get the answer to this question, ask "Why?" again and continue to do so until you are satisfied that you have actually determined the real problem, not its symptom.

L & M Thought Exercise: Cite an example of when a symptom of a problem was fixed, rather than the actual problem itself.

SEAGOING SERVIC L & M PRINCIPLE (100):
Leaders manage and managers lead.

Many people separate the functions of leadership and management into two separate distinct activities. In reality, the Leader must manage and the Manager must lead.

At the beginning of this text, you were encouraged to give some thought on the definition of leadership and management, and compare how the two were similar and different.

Most likely you defined leadership in the context of influencing others to accomplish organizational missions through communication of purpose, direction, and motivation. A Leader was the individual who did the influencing to meet his or her specified duties, directed duties, and implied duties. When you considered Managers, most likely you described a Manager as an individual engaged in the process of management. That is planning, organizing, coordinating, directing, and controlling resources such as people, physical property, money, and information to meet organizational needs.

Leaders, in order to be efficient and effective must engage in managerial duties. Repeat - Leaders, in order to be efficient and effective must engage in managerial duties. Likewise, Managers must communicate purpose, direction, assure motivation and influence others to be efficient and effective. Managers must lead.

L & M Thought Exercise: Consider the interconnectedness of the leadership and management processes.

SEAGOING SERVICE L & M PRINCIPLE (101):
Be true to your own style.

Each individual has their own personal style which is a function of their experiences, values, character, motives, learned behaviors, organizational expectations, current and previous environments, education, and skill sets.

Over time your own style will change as your body of experiences, education, and skill sets expand. Embrace the change if it is moving you as a leader and manager in a positive direction. Though we as individuals may seek to emulate the best of what we observe, we must also remain true to our own personal style. For example, if by nature you are a reserved leader and manager prone more to observation and reflection prior to leading and managing, it will seem false to yourself and others if you embrace a style different than this in a very short period of time. Of course, as indicated earlier, growth should and does occur so in the long term. Be true to your own style, while being receptive to long term growth.

L & M Thought Exercise: Reflect on your style of leadership and management. What are its strengths and weaknesses?

L & M Thought Exercise: With knowledge of the strengths and weaknesses of your style of leadership and management, what action should you initiate to improve your effectiveness?

CONCLUDING REMARKS

Congratulations on finishing your review of this text. I hope that you have gained a few insights into the leadership and management concepts which are important for professionals within the U.S. seagoing services to understand. The correct deployment of appropriate leadership and management skills is very difficult. No matter how experienced you are, you will continue to be challenged by the unique (and often frustrating) leadership and management issues within your work environment. Should you stumble in your efforts, learn what needs to be learned, dust yourself off, and forge ahead. Continue in your efforts to add value to your seagoing service, and to lead and manage your team. If you did not take the opportunity at the start of this text, please consider completing the four L & M Thought Exercises at the end of this section. I wish you the best success as you continue your service.

Ken Bryan

L & M Thought Exercise: Define Leadership and define management. How are the two similar? How are they different?

L & M Thought Exercise: What are the duties and responsibilities of a Leader?

L & M Thought Exercise: What are the duties and responsibilities of a Manager?

L & M Thought Exercise: Which is more important, the exercise of good Leadership or the exercise of good management?

Appendix A – Declaration of Independence

The Unanimous Declaration
of the Thirteen United States of America

When, in the course of human events, it becomes necessary for one people to dissolve the political bonds which have connected them with another, and to assume among the powers of the earth, the separate and equal station to which the laws of nature and of nature's God entitle them, a decent respect to the opinions of mankind requires that they should declare the causes which impel them to the separation.

We hold these truths to be self-evident, that all men are created equal, that they are endowed by their Creator with certain unalienable rights that among these are life, liberty and the pursuit of happiness. That to secure these rights, governments are instituted among men, deriving their just powers from the consent of the governed. That whenever any form of government becomes destructive to these ends, it is the right of the people to alter or to abolish it, and to institute new government, laying its foundation on such principles and organizing its powers in such form, as to them shall seem most likely to affect their safety and happiness. Prudence, indeed, will dictate that governments long established should not be changed for light and transient causes; and accordingly all experience hath shown that mankind are more disposed to suffer, while evils are sufferable, than to right themselves by abolishing the forms to which they are accustomed. But when a long train of abuses and usurpations, pursuing invariably the same object evinces a design to reduce them under absolute despotism, it is their right, it is their duty, to throw off such government, and to provide new guards for their future security. Such has been the patient sufferance of these colonies; and such is now the necessity which constrains them to alter their former systems of government. The history of the present King of

Great Britain is a history of repeated injuries and usurpations, all having in direct object the establishment of an absolute tyranny over these states. To prove this, let facts be submitted to a candid world.

He has refused his assent to laws, the most wholesome and necessary for the public good.

He has forbidden his governors to pass laws of immediate and pressing importance, unless suspended in their operation till his assent should be obtained; and when so suspended, he has utterly neglected to attend to them.

He has refused to pass other laws for the accommodation of large districts of people, unless those people would relinquish the right of representation in the legislature, a right inestimable to them and formidable to tyrants only.

He has called together legislative bodies at places unusual, uncomfortable, and distant from the depository of their public records, for the sole purpose of fatiguing them into compliance with his measures.

He has dissolved representative houses repeatedly, for opposing with manly firmness his invasions on the rights of the people.

He has refused for a long time, after such dissolutions, to cause others to be elected; whereby the legislative powers, incapable of annihilation, have returned to the people at large for their exercise; the state remaining in the meantime exposed to all the dangers of invasion from without, and convulsions within.

He has endeavored to prevent the population of these states; for that purpose obstructing the laws for naturalization of foreigners; refusing to pass others to encourage their migration hither, and raising the conditions of new appropriations of lands.

He has obstructed the administration of justice, by refusing his assent to laws for establishing judiciary powers.

He has made judges dependent on his will alone, for the tenure of their offices, and the amount and payment of their salaries.

He has erected a multitude of new offices, and sent hither swarms of officers to harass our people, and eat out their substance.

He has kept among us, in times of peace, standing armies without the consent of our legislature.

He has affected to render the military independent of and superior to civil power.

He has combined with others to subject us to a jurisdiction foreign to our constitution, and unacknowledged by our laws; giving his assent to their acts of pretended legislation:

For quartering large bodies of armed troops among us:

For protecting them, by mock trial, from punishment for any murders which they should commit on the inhabitants of these states:

For cutting off our trade with all parts of the world:

For imposing taxes on us without our consent:

For depriving us in many cases, of the benefits of trial by jury:

For transporting us beyond seas to be tried for pretended offenses:

For abolishing the free system of English laws in a neighboring province, establishing therein an arbitrary government, and enlarging its boundaries so as to render it at once an example and fit instrument for introducing the same absolute rule in these colonies:

For taking away our charters, abolishing our most valuable laws, and altering fundamentally the forms of our governments:

For suspending our own legislatures and declaring themselves invested with power to legislate for us in all cases whatsoever. He has abdicated government here, by declaring us out of his protection and waging war against us.

He has plundered our seas, ravaged our coasts, burned our towns, and destroyed the lives of our people.

He is at this time transporting large armies of foreign mercenaries to complete the works of death, desolation and tyranny, already begun with circumstances of cruelty and perfidy scarcely paralleled in the most barbarous ages, and totally unworthy the head of a civilized nation.

He has constrained our fellow citizens taken captive on the high seas to bear arms against their country, to become the executioners of their friends and brethren, or to fall themselves by their hands.

He has excited domestic insurrections amongst us, and has endeavored to bring on the inhabitants of our frontiers, the merciless Indian savages, whose known rule of warfare is undistinguished destruction of all ages, sexes and conditions.

In every stage of these oppressions we have petitioned for redress in the most humble terms: our repeated petitions have been answered only by repeated injury. A prince, whose character is thus marked by every act which may define a tyrant, is unfit to be the ruler of a free people.

Nor have we been wanting in attention to our British brethren. We have warned them from time to time of attempts by their legislature to extend an unwarrantable jurisdiction over us. We have reminded them of the circumstances of our emigration and settlement here. We have appealed to their native justice and magnanimity, and we have conjured them by the ties of our common kindred to disavow these usurpations, which, would inevitably interrupt our connections and correspondence. They too have been deaf to the voice of justice and of consanguinity. We must, therefore, acquiesce in the necessity, which denounces our separation, and hold them, as we hold the rest of mankind, enemies in war, in peace friends.

We, therefore, the representatives of the United States of America, in General Congress, assembled, appealing to the Supreme Judge of the world for the rectitude of our intentions, do, in the name, and by the authority of the good people of these colonies, solemnly publish and declare, that these united colonies are, and of right ought to be free and independent states; that they are absolved from all allegiance to the British Crown, and that all political connection between them and the state of Great Britain, is and ought to be totally dissolved; and that as free and independent states, they have full power to levy war, conclude peace, contract alliances, establish commerce, and to do all other acts and things which independent states may of right do. And for the support of this declaration, with a firm reliance on the protection of Divine Providence, we mutually pledge to each other our lives, our fortunes and our sacred honor.

Appendix B – Bill of Rights

The Bill of Rights
(Amendments I to X of the U.S. Constitution)

Amendment I: Congress shall make no law respecting an establishment of religion, or prohibiting the free exercise thereof; or abridging the freedom of speech, or of the press; or the right of the people peaceably to assemble, and to petition the Government for a redress of grievances.

Amendment II: A well regulated Militia, being necessary to the security of a free State, the right of the people to keep and bear Arms, shall not be infringed.

Amendment III: No Soldier shall, in time of peace be quartered in any house, without the consent of the Owner, nor in time of war, but in a manner to be prescribed by law.

Amendment IV: The right of the people to be secure in their persons, houses, papers, and effects, against unreasonable searches and seizures, shall not be violated, and no Warrants shall issue, but upon probable cause, supported by Oath or affirmation, and particularly describing the place to be searched, and the persons or things to be seized.

Amendment V: No person shall be held to answer for a capital, or otherwise infamous crime, unless on a presentment or indictment of a Grand Jury, except in cases arising in the land or naval forces, or in the Militia, when in actual service in time of War or public danger; nor shall any person be subject for the same offence to be twice put in jeopardy of life or limb; nor shall be compelled in any criminal case to be a witness against himself, nor be deprived of life, liberty, or property, without due process of law; nor shall private property be taken for public use, without just compensation.

Amendment VI: In all criminal prosecutions, the accused shall enjoy the right to a speedy and public trial, by an impartial jury of the State and district wherein the crime shall have been committed, which district shall have been previously ascertained by law, and to be informed of the nature and cause of the accusation; to be confronted with the witnesses against him; to have compulsory process for obtaining witnesses in his favor, and to have the Assistance of Counsel for his defense.

Amendment VII: In Suits at common law, where the value in controversy shall exceed twenty dollars, the right of trial by jury shall be preserved, and no fact tried by a jury, shall be otherwise re-examined in any Court of the United States, than according to the rules of the common law.

Amendment VIII: Excessive bail shall not be required, nor excessive fines imposed, nor cruel and unusual punishments inflicted.

Amendment IX: The enumeration in the Constitution, of certain rights, shall not be construed to deny or disparage others retained by the people.

Amendment X: The powers not delegated to the United States by the Constitution, nor prohibited by it to the States, are reserved to the States respectively, or to the people.